Geometry
Essentials
FOR
DUMMIES®

D0289780

by Mark Ryan

Geometry Essentials For Dummies®

Published by
Wiley Publishing, Inc.
111 River St.
Hoboken, NJ 07030-5774
www.wiley.com

WILEY

About the Author

A graduate of Brown University and the University of Wisconsin Law School, **Mark Ryan** has been teaching math since 1989. He runs The Math Center (www.themathcenter.com) in Winnetka, Illinois, where he teaches high school math courses, including an introduction to geometry and a workshop for parents based on a program he developed, *The 10 Habits of Highly Successful Math Students*. In high school, he twice scored a perfect 800 on the math portion of the SAT, and he not only knows mathematics but also has a gift for explaining it in plain English. He practiced law for four years before deciding he should do something he enjoys and use his natural talent for mathematics. Ryan is a member of the Authors Guild and the National Council of Teachers of Mathematics.

Geometry Essentials For Dummies is Ryan's sixth book. *Everyday Math for Everyday Life* (Grand Central Publishing) was published in 2002; *Calculus For Dummies* (Wiley), in 2003; *Calculus Workbook For Dummies* (Wiley), in 2005; *Geometry Workbook For Dummies* (Wiley), in 2006; and *Geometry For Dummies*, 2nd Edition (Wiley) in 2008. His math books have sold over a quarter of a million copies.

Also a tournament backgammon player and a skier and tennis player, Ryan lives in Chicago.

Publisher's Acknowledgments

We're proud of this book; please send us your comments at http://dummies.custhelp.com. For other comments, please contact our Customer Care Department within the U.S. at 877-762-2974, outside the U.S. at 317-572-3993, or fax 317-572-4002.

Some of the people who helped bring this book to market include the following:

Acquisitions, Editorial, and Media Development

Project Editor: Joan Friedman

Acquisitions Editor: Lindsay Sandman Lefevere

Assistant Editor: David Lutton

Technical Editors: Nancy Cozad, Amanda D. Milligan

Senior Editorial Manager: Jennifer Ehrlich

Editorial Supervisor and Reprint Editor: Carmen Krikorian

Editorial Assistant: Rachelle S. Amick

Cover Photos: © iStockphoto.com / Deanima

Cartoon: Rich Tennant (www.the5thwave.com)

Composition Services

Project Coordinator: Kristie Rees

Layout and Graphics: Carrie A. Cesavice, Corrie Socolovitch

Proofreader: Jacqui Brownstein

Indexer: Potomac Indexing, LLC

Publishing and Editorial for Consumer Dummies

Diane Graves Steele, Vice President and Publisher, Consumer Dummies

Kristin Ferguson-Wagstaffe, Product Development Director, Consumer Dummies

Ensley Eikenburg, Associate Publisher, Travel

Kelly Regan, Editorial Director, Travel

Publishing for Technology Dummies

Andy Cummings, Vice President and Publisher, Dummies Technology/General User

Composition Services

Debbie Stailey, Director of Composition Services

Contents at a Glance

Table of Contents

Introduction

Geometry is a subject full of mathematical richness and beauty. The ancient Greeks were into it big time, and it's been a mainstay in secondary education for centuries. Today, no education is complete without at least some familiarity with the fundamental principles of geometry.

But geometry is also a subject that bewilders many students because it's so unlike the math that they've done before. Geometry requires you to use deductive logic in formal proofs. This process involves a special type of verbal and mathematical reasoning that's new to many students. The subject also involves working with two- and three-dimensional shapes. The spatial reasoning required for this is another thing that makes geometry different and challenging.

Geometry Essentials For Dummies can be a big help to you if you've hit the geometry wall. Or if you're a first-time student of geometry, it can prevent you from hitting the wall in the first place. When the world of geometry opens up to you and things start to click, you may come to really appreciate this topic, which has fascinated people for millennia.

About This Book

Geometry Essentials For Dummies covers all the principles and formulas you need to analyze two- and three-dimensional shapes, and it gives you the skills and strategies you need to write geometry proofs.

My approach throughout is to explain geometry in plain English with a minimum of technical jargon. Plain English suffices for geometry because its principles, for the most part, are accessible with your common sense. I see no reason to obscure geometry concepts behind a lot of fancy-pants mathematical mumbo-jumbo. I prefer a street-smart approach.

This book, like all *For Dummies* books, is a reference, not a tutorial. The basic idea is that the chapters stand on their own as much as possible. So you don't have to read this book cover to cover — although, of course, you might want to.

Conventions Used in This Book

Geometry Essentials For Dummies follows certain conventions that keep the text consistent:

- ✔ Variables and names of points are in *italics.*

- ✔ Important math terms are often in *italics* and are defined when necessary. Italics are also sometimes used for emphasis.

- ✔ Important terms may be **bolded** when they appear as keywords within a bulleted list. I also use bold for the instructions in many-step processes.

- ✔ As in most geometry books, figures are not necessarily drawn to scale — though most of them are.

Foolish Assumptions

As I wrote this book, here's what I assumed about you:

- ✔ You're a high school student (or perhaps a junior high student) currently taking a standard high school–level geometry course, or . . .

- ✔ You're a parent of a geometry student, and you'd like to understand the fundamentals of geometry so you can help your child do his or her homework and prepare for quizzes and tests, or . . .

- ✔ You're anyone who wants to refresh your recollection of the geometry you studied years ago or wants to explore geometry for the first time.

- ✔ You remember some basic algebra. The good news is that you need very little algebra for doing geometry — but you do need some. In the problems that do involve algebra, I try to lay out all the solutions step by step.

Icons Used in This Book

Next to this icon are definitions of geometry terms, explanations of geometry principles, and a few other things you should remember as you work through the book.

This icon highlights shortcuts, memory devices, strategies, and so on.

Ignore these icons, and you may end up doing lots of extra work or getting the wrong answer or both. Read carefully when you see the bomb with the burning fuse!

This icon identifies the theorems and postulates — little mathematical truths — that you use to form the logical arguments in geometry proofs.

Where to Go from Here

If you're a geometry beginner, you should probably start with Chapter 1 and work your way through the book in order, but if you already know a fair amount of the subject, feel free to skip around. For instance, if you need to know about quadrilaterals, check out Chapter 6. Or if you already have a good handle on geometry proof basics, you may want to dive into the more advanced proofs in Chapter 5.

And from there, naturally, you can go

- ✔ To the head of the class
- ✔ To Go to collect $200
- ✔ To chill out
- ✔ To explore strange new worlds, to seek out new life and new civilizations, to boldly go where no man (or woman) has gone before.

If you're still reading this, what are you waiting for? Go take your first steps into the wonderful world of geometry!

The 5th Wave

By Rich Tennant

"I hear you think you got all the angles figured. Well, maybe you do and maybe you don't. Maybe the ratios of the lengths of corresponding sides of an equiangular right-angled triangle are equal, then again, maybe they're not — let's see your equations."

Chapter 1

An Overview of Geometry

Studying geometry is sort of a Dr. Jekyll-and-Mr. Hyde thing. You have the ordinary geometry of shapes (the Dr. Jekyll part) and the strange world of geometry proofs (the Mr. Hyde part).

Every day, you see various shapes all around you (triangles, rectangles, boxes, circles, balls, and so on), and you're probably already familiar with some of their properties: area, perimeter, and volume, for example. In this book, you discover much more about these basic properties and then explore more advanced geometric ideas about shapes.

Geometry proofs are an entirely different sort of animal. They involve shapes, but instead of doing something straightforward like calculating the area of a shape, you have to come up with a mathematical argument that proves something about a shape. This process requires not only mathematical skills but verbal skills and logical deduction skills as well, and for this reason, proofs trip up many, many students. If you're one of these people and have already started singing the geometry-proof blues, you might even describe proofs — like Mr. Hyde — as monstrous. But I'm confident that, with the help of this book, you'll have no trouble taming them.

The Geometry of Shapes

Have you ever reflected on the fact that you're literally surrounded by shapes? Look around. The rays of the sun are — what else? — rays. The book in your hands has a shape, every table and chair has a shape, every wall has an area, and every container has a shape and a volume; most picture frames are rectangles, DVDs are circles, soup cans are cylinders, and so on.

One-dimensional shapes

There aren't many shapes you can make if you're limited to one dimension. You've got your lines, your segments, and your rays. That's about it. On to something more interesting.

Two-dimensional shapes

As you probably know, two-dimensional shapes are flat things like triangles, circles, squares, rectangles, and pentagons. The two most common characteristics you study about 2-D shapes are their area and perimeter. I devote many chapters in this book to triangles and *quadrilaterals* (shapes with four sides); I give less space to shapes that have more sides, like pentagons and hexagons. Then there are the shapes with curved sides: The only curved 2-D shape I discuss is the circle.

Three-dimensional shapes

In this book, you work with prisms (a box is one example), cylinders, pyramids, cones, and spheres. The two major characteristics of these 3-D shapes are their *surface area* and *volume*. These two concepts come up frequently in the real world; examples include the amount of wrapping paper you need to wrap a gift box (a surface area problem) and the volume of water in a backyard pool (a volume problem).

Geometry Proofs

A *geometry proof* — like any mathematical proof — is an argument that begins with known facts, proceeds from there through a series of logical deductions, and ends with the thing you're trying to prove. Here's a very simple example using the line segments in Figure 1-1.

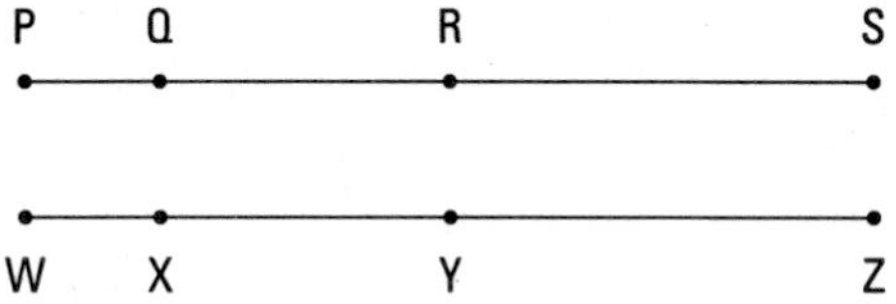

P   Q   R   S

W   X   Y   Z

Figure 1-1: *PS* and *WZ*, each made up of three pieces.

For this proof, you're told that segment $\overline{PS}$ is *congruent to* (the same length as) segment $\overline{WZ}$, that $\overline{PQ}$ is congruent to $\overline{WX}$, and that $\overline{QR}$ is congruent to $\overline{XY}$. You have to prove that $\overline{RS}$ is congruent to $\overline{YZ}$.

Now, you may be thinking, "That's obvious — if $\overline{PS}$ is the same length as $\overline{WZ}$ and both segments contain these equal short pieces and the equal medium pieces, then the longer third pieces have to be equal as well." And you'd be right. But that's not how the proof game is played. You have to spell out every little step in your thinking. Here's the whole chain of logical deductions:

1. $\overline{PS} \cong \overline{WZ}$ (this is given).

2. $\overline{PQ} \cong \overline{WX}$ and $\overline{QR} \cong \overline{XY}$ (these facts are also given).

3. Therefore, $\overline{PR} \cong \overline{WY}$ (because if you add equal things to equal things, you get equal totals).

4. Therefore, $\overline{RS} \cong \overline{YZ}$ (because if you start with equal segments, the whole segments $\overline{PS}$ and $\overline{WZ}$, and take away equal parts of them, $\overline{PR}$ and $\overline{WY}$, the parts that are left must be equal).

Am I Ever Going to Use This?

You'll likely have plenty of opportunities to use your knowledge about the geometry of shapes. What about geometry proofs? Not so much.

When you'll use your knowledge of shapes

Shapes are everywhere, so every educated person should have a working knowledge of shapes and their properties. If you have to buy fertilizer or grass seed for your lawn, you should know something about area. You might want to understand the volume measurements in cooking recipes, or you may want to help a child with an art or science project that involves geometry. You certainly need to understand something about geometry to build some shelves or a backyard deck. And after finishing your work, you might be hungry — a grasp of how area works can come in handy when you're ordering pizza: a 20-inch pizza is four, not two, times as big as a 10-incher. There's no end to the list of geometry problems that come up in everyday life.

When you'll use your knowledge of proofs

Will you ever use your knowledge of geometry proofs? I'll give you a politically correct answer and a politically incorrect one. Take your pick.

First, the politically correct answer (which is also *actually* correct). Granted, it's extremely unlikely that you'll ever have occasion to do a single geometry proof outside of a high school math course. However, doing geometry proofs teaches you important lessons that you can apply to nonmathematical arguments. Proofs teach you . . .

- ✔ Not to assume things are true just because they seem true

- ✔ To carefully explain each step in an argument even if you think it should be obvious to everyone

- ✔ To search for holes in your arguments

- ✔ Not to jump to conclusions

In general, proofs teach you to be disciplined and rigorous in your thinking and in communicating your thoughts.

If you don't buy that PC stuff, I'm sure you'll get this politically incorrect answer: Okay, so you're never going to use geometry proofs, but you want to get a decent grade in geometry, right? So you might as well pay attention in class (what else is there to do, anyway?), do your homework, and use the hints, tips, and strategies I give you in this book. They'll make your life much easier. Promise.

Getting Down with Definitions

The study of geometry begins with the definitions of the five simplest geometric objects: point, line, segment, ray, and angle. And I throw in two extra definitions for you (plane and 3-D space) for no extra charge.

> ✔ **Point:** A point is like a dot except that it has no size at all. A point is zero-dimensional, with no height, length, or width, but you draw it as a dot, anyway. You name a point with a single uppercase letter, as with points *A*, *D*, and *T* in Figure 1-2.

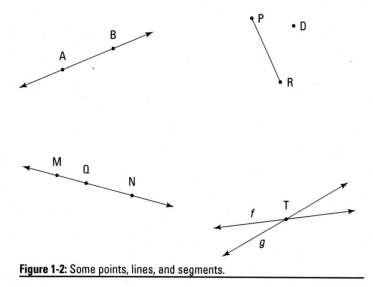

Figure 1-2: Some points, lines, and segments.

- **Line:** A line is like a thin, straight wire (although really it's infinitely thin — or better yet, it has no width at all). Lines have length, so they're one-dimensional. Remember that a line goes on forever in both directions, which is why you use the little double-headed arrow as in $\overleftrightarrow{AB}$ (read as *line AB*).

  Check out Figure 1-2 again. Lines are usually named using any two points on the line, with the letters in any order. So $\overleftrightarrow{MQ}$ is the same line as $\overleftrightarrow{QM}$, $\overleftrightarrow{MN}$, $\overleftrightarrow{NM}$, $\overleftrightarrow{QN}$, and $\overleftrightarrow{NQ}$. Occasionally, lines are named with a single, italicized, lowercase letter, such as lines *f* and *g*.

- **Line segment (or just segment):** A segment is a section of a line that has two endpoints. See Figure 1-2 yet again. If a segment goes from *P* to *R*, you call it *segment PR* and write it as $\overline{PR}$. You can also switch the order of the letters and call it $\overline{RP}$. Segments can also appear within lines, as in $\overline{MN}$.

- **Ray:** A ray is a section of a line (kind of like half a line) that has one endpoint and goes on forever in the other direction. If its endpoint is point *K* and it goes through point *S* and then past *S* forever, you call it *ray KS* and write $\overrightarrow{KS}$. See Figure 1-3. Note that a ray like $\overrightarrow{AB}$ can also be called $\overrightarrow{AC}$ because either way, you start at *A* and go forever past *B* and *C*. $\overrightarrow{BC}$, however, is a different ray.

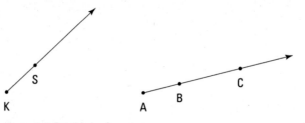

Figure 1-3: Catching a few rays.

- **Angle:** Two rays with the same endpoint form an angle. Each ray is a *side* of the angle, and the common endpoint is the angle's *vertex*. You can name an angle using its vertex alone or three points (first, a point on one ray, then the vertex, and then a point on the other ray).

Check out Figure 1-4. Rays $\overrightarrow{PQ}$ and $\overrightarrow{PR}$ form the sides of an angle, with point P as the vertex. You can call the angle $\angle P$, $\angle RPQ$, or $\angle QPR$. Angles can also be named with numbers, such as the angle on the right in the figure, which you can call $\angle 4$. The number is just another way of naming the angle, and it has nothing to do with the size of the angle. The angle on the right also illustrates the *interior* and *exterior* of an angle.

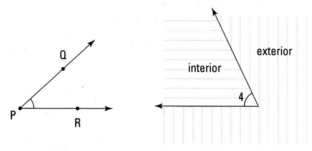

Figure 1-4: Some angles and their parts.

✔ **Plane:** A plane is like a perfectly flat sheet of paper except that it has no thickness whatsoever and it goes on forever in all directions. You might say it's infinitely thin and has an infinite length and an infinite width. Because it has length and width but no height, it's two-dimensional. Planes are named with a single, italicized, lowercase letter or sometimes with the name of a figure (a rectangle, for example) that lies in the plane.

✔ **3-D (three-dimensional) space:** 3-D space is everywhere — all of space in every direction. First, picture an infinitely big map that goes forever to the north, south, east, and west. That's a two-dimensional plane. Then, to get 3-D space from this map, add the third dimension by going up and down forever.

A Few Points on Points

There isn't much that can be said about points. They have no features, and each one is the same as every other. Various *groups* of points, however, do merit an explanation:

✔ **Collinear points:** See the word *line* in *collinear?* Collinear points are points that lie on a line. Any two points are always collinear because you can always connect them with a straight line. Three or more points can be collinear, but they don't have to be.

✔ **Non-collinear points:** Non-collinear points are three or more points that don't all lie on the same line.

✔ **Coplanar points:** A group of points that lie in the same plane are coplanar. Any two or three points are always coplanar. Four or more points might or might not be coplanar.

Look at Figure 1-5, which shows coplanar points *A, B, C,* and *D.* In the box on the right, there are many sets of coplanar points. Points *P, Q, X,* and *W,* for example, are coplanar; the plane that contains them is the left side of the box. Note that points *Q, X, S,* and *Z* are also coplanar even though the plane that contains them isn't shown; it slices the box in half diagonally.

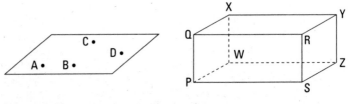

Figure 1-5: Coplanar and non-coplanar points.

✔ **Non-coplanar points:** A group of points that don't all lie in the same plane are non-coplanar. In Figure 1-5, points *P, Q, X,* and *Y* are non-coplanar. The top of the box contains *Q, X,* and *Y,* and the left side contains *P, Q,* and *X,* but no plane contains all four points.

Lines, Segments, and Rays

In this section, I describe different types of lines (or segments or rays) or pairs of lines (or segments or rays) based on the direction they're pointing or how they relate to each other.

Horizontal and vertical lines

Defining *horizontal* and *vertical* may seem a bit pointless because you probably already know what the terms mean. But, hey, this is a math book, and math books are supposed to define terms. Who am I to question this tradition?

✔ **Horizontal lines, segments, or rays:** Horizontal lines, segments, and rays go straight across, left and right, not up or down at all — you know, like the horizon.

✔ **Vertical lines, segments, or rays:** Lines or parts of a line that go straight up and down are vertical. (Shocking!)

Doubling up with pairs of lines

In this section, I give you five terms that describe pairs of lines. The first four are about coplanar lines — you use these a lot. The fifth term describes non-coplanar lines. This term comes up only in 3-D problems, so you won't need it much.

Coplanar lines

Coplanar lines are lines in the same plane. Here are some ways coplanar lines may interact:

✔ **Parallel lines, segments, or rays:** Lines that run in the same direction and never cross (like two railroad tracks) are called parallel. Segments and rays are parallel if the lines that contain them are parallel. If $\overleftrightarrow{AB}$ is parallel to $\overleftrightarrow{CD}$, you write $\overleftrightarrow{AB} \parallel \overleftrightarrow{CD}$.

✔ **Intersecting lines, segments, or rays:** Lines, rays, or segments that cross or touch are intersecting.

• **Perpendicular lines, segments, or rays:** Lines, segments, or rays that intersect at right angles are perpendicular. If $\overline{PQ}$ is perpendicular to $\overline{RS}$, you write $\overline{PQ} \perp \overline{RS}$. See Figure 1-6. The little boxes in the corners of the angles indicate right angles.

• **Oblique lines, segments, or rays:** Lines or segments or rays that intersect at any angle other than 90° are called *oblique*. See Figure 1-6.

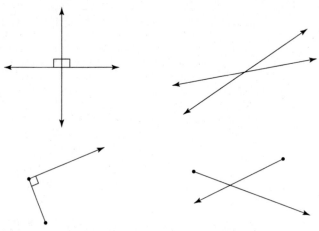

Figure 1-6: Perpendicular and oblique lines, rays, and segments.

Non-coplanar lines

Non-coplanar lines are lines that cannot be contained in a single plane.

Skew lines, segments, or rays: Lines that don't lie in the same plane are called skew lines — *skew* simply means *non-coplanar.* Or you can say that skew lines are lines that are neither parallel nor intersecting.

Investigating the Plane Facts

Here are two terms for a pair of planes (see Figure 1-7):

- ✔ **Parallel planes:** Parallel planes are planes that never cross. The ceiling of a room (assuming it's flat) and the floor are parallel planes (though true planes extend forever).

- ✔ **Intersecting planes:** Hold onto your hat — intersecting planes are planes that cross, or intersect. When planes intersect, the place where they cross forms a line. The floor and a wall of a room are intersecting planes, and where the floor meets the wall is the line of intersection of the two planes.

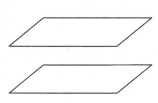
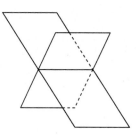

Parallel planes   Intersecting planes

Figure 1-7: Parallel and intersecting planes.

Everybody's Got an Angle

Angles are one of the basic building blocks of triangles and other polygons. Angles appear on virtually every page of every geometry book, so you gotta get up to speed about them — no ifs, ands, or buts.

Five types of angles

Check out the five angle definitions and see Figure 1-8:

- **Acute angle:** An acute angle is less than 90°. Think "a-*cute* little angle."

- **Right angle:** A right angle is a 90° angle. Right angles should be familiar to you from the corners of picture frames, tabletops, boxes, and books, and all kinds of other things that show up in everyday life.

- **Obtuse angle:** An obtuse angle has a measure greater than 90°.

- **Straight angle:** A straight angle has a measure of 180°; it looks just like a line with a point on it.

- **Reflex angle:** A reflex angle has a measure of more than 180°. Basically, a reflex angle is just the other side of an ordinary angle.

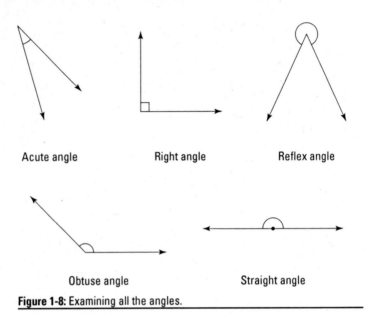

Figure 1-8: Examining all the angles.

Angle pairs

The needy angles in this section have to be in a relationship with another angle for these definitions to mean anything.

↘ **Adjacent angles:** Adjacent angles are neighboring angles that have the same vertex and that share a side; also, neither angle can be inside the other. I realize that's quite a mouthful. This very simple idea is kind of a pain to define, so just check out Figure 1-9. ∠*BAC* and ∠*CAD* are adjacent, as are ∠1 and ∠2. However, neither ∠1 nor ∠2 is adjacent to ∠*XYZ* because they're both inside ∠*XYZ*. None of the unnamed angles to the right are adjacent because they either don't share a vertex or don't share a side.

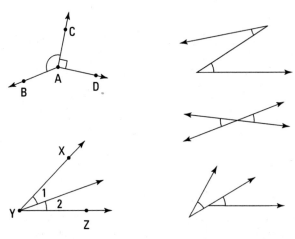

Figure 1-9: Adjacent and non-adjacent angles.

> ✔ **Complementary angles:** Two angles that add up to 90° are complementary. They can be adjacent angles but don't have to be. In Figure 1-10, adjacent angles ∠1 and ∠2 are complementary because they make a right angle; ∠*P* and ∠*Q* are complementary because they add up to 90°.

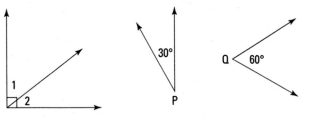

Figure 1-10: Complementary angles can join forces to form a right angle.

> ✔ **Supplementary angles:** Two angles that add up to 180° are supplementary. They may or may not be adjacent angles. In Figure 1-11, ∠1 and ∠2, or the two right angles, are supplementary because they form a straight angle. Such angle pairs are called a *linear pair*. Angles *A* and *Z* are supplementary because they add up to 180°.

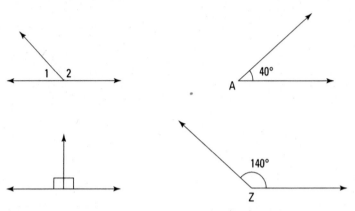

Figure 1-11: Together, supplementary angles can form a straight line.

> ✔ **Vertical angles:** When two intersecting lines form an X,
> the angles on the opposite sides of the X are called verti-
> cal angles. Two vertical angles are always congruent. By
> the way, the *vertical* in *vertical angles* has nothing to do
> with the up-and-down meaning of the word vertical.

Bisection and Trisection

For all you fans of bicycles and tricycles and bifocals and
trifocals — not to mention the biathlon and the triathlon,
bifurcation and trifurcation, and bipartition and tripartition —
you're really going to love this section on bisection and
trisection: cutting something into two or three equal parts.

Segments

Segment *bisection*, the related term *midpoint*, and segment
trisection are pretty simple ideas.

> ✔ **Segment bisection:** A point, segment, ray, or line that
> divides a segment into two congruent segments *bisects*
> the segment.

> ✔ **Midpoint:** The point where a segment is bisected is
> called the *midpoint*; the midpoint cuts the segment into
> two congruent parts.

> ✔ **Segment trisection:** Two things (points, segments, rays, or lines) that divide a segment into three congruent segments *trisect* the segment.

WARNING!

Students often make the mistake of thinking that *divide* means to *bisect,* or cut exactly in half. This error is understandable because when you do ordinary division with numbers, you are, in a sense, dividing the larger number into equal parts (24 ÷ 2 = 12 because 12 + 12 = 24). But in geometry, to *divide* something just means to cut it into parts of any size, equal or unequal. *Bisect* and *trisect,* of course, *do* mean to cut into exactly equal parts.

REMEMBER

Angles

Brace yourself for a shocker: The terms *bisecting* and *trisecting* mean the same thing for angles as they do for segments!

> ✔ **Angle bisection:** A ray that cuts an angle into two congruent angles *bisects* the angle. The ray is called the *angle bisector.*

> ✔ **Angle trisection:** Two rays that divide an angle into three congruent angles *trisect* the angle. These rays are called *angle trisectors.*

Take a stab at this problem: In Figure 1-12, $\overrightarrow{TP}$ bisects $\angle STL$, which equals $(12x - 24)°$; $\overrightarrow{TL}$ bisects $\angle PTI$, which equals $(8x)°$. Is $\angle STI$ trisected, and what is its measure?

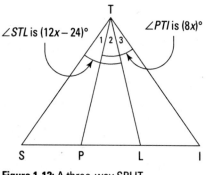

Figure 1-12: A three-way SPLIT.

Nothing to it. First, yes, ∠*STI* is trisected. You know this because ∠*STL* is bisected, so ∠1 must equal ∠2. And because ∠*PTI* is bisected, ∠2 equals ∠3. Thus, all three angles must be equal, and that means ∠*STI* is trisected.

Now find the measure of ∠*STI*. Because ∠*STL* — which measures $(12x - 24)°$ — is bisected, ∠2 must be half its size, or $(6x - 12)°$. And because ∠*PTI* is bisected, ∠2 must also be half the size of ∠*PTI* — that's half of $(8x)°$, or $(4x)°$. Because ∠2 equals both $(6x - 12)°$ and $(4x)°$, you set those expressions equal to each other and solve for *x*:

$$6x - 12 = 4x$$
$$2x = 12$$
$$x = 6$$

Then just plug $x = 6$ into, say, $(4x)°$, which gives you $4 \cdot 6$, or $24°$ for ∠2. Angle *STI* is three times that, or $72°$. That does it.

When rays trisect an angle of a triangle, the opposite side of the triangle is *never* trisected by these rays. (It might be close, but it's never exactly trisected.) In Figure 1-12, for instance, because ∠*STI* is trisected, $\overline{SI}$ is definitely *not* trisected by points *P* and *L*. This warning also works in reverse. Assuming for the sake of argument that *P* and *L* *did* trisect $\overline{SI}$, you would know that ∠*STI* is definitely *not* trisected.

Chapter 2

Geometry Proof Starter Kit

- -

In This Chapter

▶ Introducing if-then logic

▶ Theorizing about theorems and defining definitions

▶ Understanding the complementary and supplementary angle theorems

- -

*T*raditional two-column geometry proofs are arguably the most important — and for many students, the most difficult — topic in a standard high school geometry course. In this chapter, I aim to declaw them! First, I give you a schematic drawing that shows you all the elements of a two-column proof and where they go. Then, I explain how you prove something using a deductive argument. After that, you'll be ready for a starter kit of your first 18 theorems along with some proofs that illustrate how to use them. Before you know it, you'll be a proof-writing crackerjack!

The Lay of the (Proof) Land

A geometry proof involves a geometric diagram of some sort. You're told some things that are true about the diagram (the *givens*), and you're asked to prove that something else is true (the *prove* statement). Every proof proceeds as follows:

1. You begin with one or more of the given facts about the diagram.

2. You state something that follows from the given fact or facts; then you state something that follows from that; then, something that follows from that; and so on. Each deduction leads to the next.

3. You end by making your final deduction — the fact you're trying to prove.

Geometry proofs contain the following elements. Figure 2-1 shows how these elements fit together.

- ✔ **The diagram:** The shape or shapes in the diagram are the subject matter of the proof. Your goal is to prove some fact about the diagram (for example, that two triangles in the diagram are congruent).

- ✔ **The givens:** The givens are true facts about the diagram that you build upon to reach your goal, the *prove* statement. You always begin a proof with one of the givens, putting it in line 1 of the statement column.

- ✔ **The *prove* statement:** The *prove* statement is the fact about the diagram that you must establish with your chain of logical deductions. It always goes in the last line of the statement column.

- ✔ **The statement column:** In the statement column, you put all the given facts, the facts that you deduce, and in the final line, the *prove* statement. In this column, you put *specific* facts about *specific* geometric objects, such as $\angle ABD \cong \angle CBD$.

- ✔ **The reason column:** In this column, you put the justification for each statement that you make. In this column, you write *general* rules such as *If an angle is bisected, then it's divided into two congruent parts.* You do not give the names of specific objects.

Reasoning with If-Then Logic

Every geometry proof is a sequence of logical deductions. You write one of the given facts as statement 1. Then, for statement 2, you put something that follows from statement 1 and write your justification for that in the reason column. Then you proceed to statement 3, and so on, till you get to the *prove* statement. The way you get from statement 1 to statement 2, from statement 2 to statement 3, and so on is by using *if-then* logic.

The givens are located after the word "Given." Will wonders never cease!

This, obviously, is the diagram.

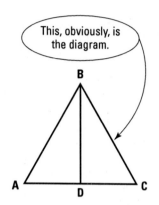

Given: $\overrightarrow{BD}$ bisects $\angle ABC$

∠A is complementary to ∠ABD
∠C is complementary to ∠CBD

Prove: $\angle A \cong \angle C$ Here's your goal, the *prove* statement.

Statements	Reasons
(You put your statements in this column.)	(Your reasons go in this column.)
1) $\overrightarrow{BD}$ bisects $\angle ABC$ (One of the givens always goes on line 1.)	1) **Given.** ("Given" is always the justification for Statement #1.)
2) (Other givens and facts you deduce 3) go here.) . .	2) (Justifications for your statements 3) go here.) . .
Last) $\angle A \cong \angle C$ (The *prove* statement always goes on the last line.)	**Last) Complements of congruent angles are congruent.** (The justification for the *prove* statement goes here.)

Figure 2-1: Anatomy of a geometry proof.

If-then chains of logic

A two-column geometry proof is in essence a logical argument or a chain of logical deductions, like

1. If I study, then I'll get good grades.
2. If I get good grades, then I'll get into a good college.
3. If I get into a good college, then I'll become a babe/guy magnet.
4. (And so on . . .)

Note that each of these steps is a sentence with an *if* clause and a *then* clause.

Here's an example of a two-column proof from everyday life. Say you've got a Dalmatian named Spot, and you want to prove that he's a mammal. Figure 2-2 shows the proof. On the first line of the statement column, you put down the given fact that Spot is a Dalmatian, and you write *Given* in the reason column. Then, in statement 2, you put down a new fact that you deduce from statement 1 — namely, *Spot is a dog*. In reason 2, you justify or defend that claim with the reason *If something is a Dalmatian, then it's a dog.*

Statements (or Conclusions)	Reasons (or Justifications)
1) Spot is a Dalmatian	1) Given.
2) Spot is a dog	2) If something is a Dalmatian, then it's a dog.
3) Spot is a mammal	3) If something is a dog, then it's a mammal.

Figure 2-2: Proving that Spot is a mammal.

Here's a good way to think about how reasons work. Say you write a reason like *If something is a Dalmatian, then it's a dog.* You can think of the word *if* as meaning *because I already know,* and you can think of the word *then* as meaning *I can now deduce.* So basically, the second reason in Figure 2-2 means that because you already know that Spot is a Dalmatian, you can deduce or conclude that Spot is a dog.

Continuing with the proof, in statement 3, you write something that you can deduce from statement 2, namely that Spot is a mammal. For reason 3, you write your justification for statement 3: *If something is a dog, then it's a mammal.* Every geometry proof has this same basic structure.

Definitions, theorems, and postulates

Definitions, theorems, and postulates are the building blocks of geometry proofs. With very few exceptions, every justification in the reason column is one of these three things. Look back at Figure 2-2. If that had been a geometry proof instead of a dog proof, the reason column would contain *if-then* definitions, theorems, and postulates about geometry instead of *if-then* ideas about dogs. Here's the lowdown on definitions, theorems, and postulates.

Using definitions in the reason column

Definition: I'm sure you know what a definition is — it defines or explains what a term means. Here's an example: "A *midpoint* divides a segment into two congruent parts."

You can write all definitions in *if-then* form in either direction: "If a point is a midpoint of a segment, then it divides that segment into two congruent parts" or "If a point divides a segment into two congruent parts, then it's the midpoint of that segment." Figure 2-3 shows you how to use both versions of the midpoint definition in a proof. Make sure you understand how to use the two versions.

When you have to choose between these two versions of the midpoint definition, remember that you can think of the word *if* as meaning *because I already know* and the word *then* as meaning *I can now deduce.* For example, for reason 2 in the first proof in Figure 2-3, you choose the version that goes, "*If* a point is the midpoint of a segment, *then* it divides the segment into two congruent parts," because you already know that *M* is the midpoint of (because it's given) and from that given fact you can deduce that $\overline{AM} \cong \overline{MB}$.

Using theorems and postulates as reasons

Both theorems and postulates are statements of geometrical truth, such as *All right angles are congruent* or *All radii of a circle are congruent.* The difference between postulates and theorems is that postulates are assumed to be true, but theorems must be proven to be true based on postulates and/or already-proven theorems. It's a fine distinction, and if I were you, I wouldn't sweat it.

The following mini proofs use this figure:

A ———•——— M ———•——— B

For the first mini proof, it's *given* that *M* is the midpoint of $\overline{AB}$, and you need to *prove* that $\overline{AM} \cong \overline{MB}$.

Statements (or Conclusions)	Reasons (or Justifications)
1) *M* is the midpoint of $\overline{AB}$	1) Given.
2) $\overline{AM} \cong \overline{MB}$	2) *If* a point is the midpoint of a segment, *then* it divides the segment into two congruent parts.

For the second mini proof, it's *given* $\overline{AM} \cong \overline{MB}$, and you need to *prove* that *M* is the midpoint of $\overline{AB}$.

Statements (or Conclusions)	Reasons (or Justifications)
1) $\overline{AM} \cong \overline{MB}$	1) Given.
2) *M* is the midpoint of $\overline{AB}$	2) *If* a point divides a segment into two congruent parts, *then* it's the midpoint of the segment.

Figure 2-3: Double duty — using both versions of the midpoint definition in the reason column.

Written in *if-then* form, the theorem *All right angles are congruent* would read, "If two angles are right angles, then they are congruent." Unlike definitions, theorems (and postulates) are generally *not* reversible. For example, if you reverse this right-angle theorem, you get a false statement: "If two angles are congruent, then they are right angles." (If a theorem works in both directions, you'll get a separate theorem for each version.) Theorems and postulates are used in the reason column exactly like definitions are used.

When you're doing your first proofs, or later if you're struggling with a difficult one, it's very helpful to write your reasons (definitions, theorems, and postulates) in *if-then* form. When you use *if-then* form, the logical structure of the proof is easier to follow. After you become a proof expert, you can abbreviate your reasons in non-*if-then* form or simply list the name of the definition, theorem, or postulate.

Bubble logic

I like to add bubbles and arrows to a proof solution to show the connections between the statements and the reasons. You won't be asked to do this when you solve a proof; it's just a way to help you understand how proofs work. Figure 2-4 shows the Spot-the-dog proof from Figure 2-2, this time with bubbles and arrows that show how the logic flows through the proof.

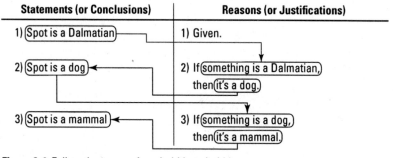

Figure 2-4: Follow the arrows from bubble to bubble.

Follow the arrows from bubble to bubble, and follow my tips to stay out of trouble! (It's because of high poetry like this that they pay me the big bucks.)

This is huge! Take heed. In a two-column proof,

- ✔ The idea (or ideas) in the *if* clause of each reason must come from the statement column somewhere *above* the reason.

- ✔ The single idea in the *then* clause of each reason must match the idea in the statement on the *same line* as the reason.

Complementary and Supplementary Angles

Ready for your first theorems?

✔ **Complements of the same angle are congruent.** If two angles are each complementary to a third angle, then they're congruent to each other. (This theorem involves three total angles.)

✔ **Complements of congruent angles are congruent.** If two angles are complementary to two other congruent angles, then they're congruent (four total angles).

✔ **Supplements of the same angle are congruent.** If two angles are each supplementary to a third angle, then they're congruent to each other. (Three-angle version.)

✔ **Supplements of congruent angles are congruent.** If two angles are supplementary to two other congruent angles, then they're congruent. (Four-angle version.)

These four theorems, as well as the upcoming addition and subtraction theorems and the transitivity theorems, come in pairs: One of the theorems involves *three* segments or angles, and the other, which is based on the same idea, involves *four* segments or angles. When doing a proof, note whether the relevant part of the proof diagram contains three or four segments or angles to determine whether to use a three- or four-thing version.

Take a look at one of the complementary-angle theorems and one of the supplementary-angle theorems in action:

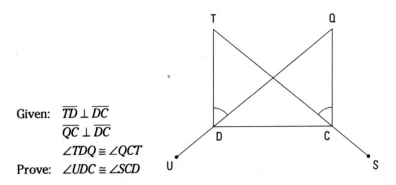

Given: $\overline{TD} \perp \overline{DC}$
$\overline{QC} \perp \overline{DC}$
$\angle TDQ \cong \angle QCT$
Prove: $\angle UDC \cong \angle SCD$

Before trying to write out a formal, two-column proof, it's often a good idea to think through a seat-of-the-pants argument about why the *prove* statement has to be true. I call this argument a *game plan*. Game plans are especially helpful for longer proofs.

When working through a game plan, you may find it helpful to make up arbitrary sizes for segments and angles in the proof. You can do this for segments and angles in the givens and, sometimes, for unmentioned segments and angles. You should *not*, however, make up sizes for things that you're trying to show are congruent.

Game plan: In this proof, for example, you might say to yourself, "Let's see . . . Because of the given perpendicular segments, I have two right angles. Next, the other given tells me that $\angle TDQ \cong \angle QCT$. If they were both 50°, $\angle QDC$ and $\angle TCD$ would both be 40°, and then $\angle UDC$ and $\angle SCD$ would both have to be 140° (because a straight line is 180°)." That does it.

Statements	Reasons
1) $\overline{TD} \perp \overline{DC}$ $\overline{QC} \perp \overline{DC}$	1) Given. (Why would they tell you this? See reason 2.)
2) $\angle TDC$ is a right angle $\angle QCD$ is a right angle	2) If segments are perpendicular, then they form right angles (definition of perpendicular).
3) $\angle CDQ$ is complementary to $\angle TDQ$, $\angle DCT$ is complementary to $\angle QCT$	3) If two angles form a right angle, then they're complementary (definition of complementary angles).
4) $\angle TDQ \cong \angle QCT$	4) Given.
5) $\angle CDQ \cong \angle DCT$	5) If two angles are complementary to two other congruent angles, then they're congruent.
6) $\angle UDQ$ is a straight angle $\angle SCT$ is a straight angle	6) Assumed from diagram.
7) $\angle UDC$ is supplementary to $\angle CDQ$, $\angle SCD$ is supplementary to $\angle DCT$	7) If two angles form a straight angle, then they're supplementary (definition of supplementary angles).
8) $\angle UDC \cong \angle SCD$	8) If two angles are supplementary to two other congruent angles, then they're congruent.

Addition and Subtraction

In this section, I give you eight simple theorems: four about adding or subtracting segments and four (that work exactly the same way) about adding or subtracting angles.

Addition theorems

✔ **Segment addition (three total segments):** If a segment is added to two congruent segments, then the sums are congruent.

✔ **Angle addition (three total angles):** If an angle is added to two congruent angles, then the sums are congruent.

After you're comfortable with proofs and know your theorems well, you can abbreviate these theorems as *segment addition* or *angle addition* or simply *addition;* however, when you're starting out, writing the theorems out in full is a good idea.

Figure 2-5 shows you how these two theorems work.

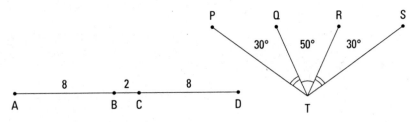

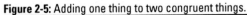
Figure 2-5: Adding one thing to two congruent things.

If you add $\overline{BC}$ to the congruent segments $\overline{AB}$ and $\overline{CD}$, the sums, namely $\overline{AC}$ and $\overline{BD}$, are congruent. In other words, $8 + 2 = 8 + 2$. Extraordinary!

And if you add $\angle QTR$ to congruent angles $\angle PTQ$ and $\angle RTS$, the sums, $\angle PTR$ and $\angle QTS$, will be congruent: $30° + 50° = 30° + 50°$. Brilliant!

✔ **Segment addition (four total segments):** If two congruent segments are added to two other congruent segments, then the sums are congruent.

✔ **Angle addition (four total angles):** If two congruent angles are added to two other congruent angles, then the sums are congruent.

Check out Figure 2-6, which illustrates these theorems.

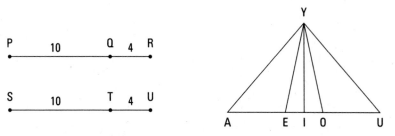

Figure 2-6: Adding congruent things to congruent things.

If $\overline{PQ}$ and $\overline{ST}$ are congruent and $\overline{QR}$ and $\overline{TU}$ are congruent, then $\overline{PR}$ is obviously congruent to $\overline{SU}$, right?

And if $\angle AYE \cong \angle UYO$ (say they're both 40°) and $\angle EYI \cong \angle OYI$ (say they're both 20°), then $\angle AYI \cong \angle UYI$ (they'd both be 60°).

Now for a proof that uses segment addition:

Given: $\overline{MD} \cong \overline{VI}$
$\overline{DX} \cong \overline{CV}$

Prove: $\overline{MC} \cong \overline{XI}$

Impress me: What year is MDXCVI?

Really impress me: What famous mathematician (who made a major breakthrough in geometry) was born in this year?

I've put what amounts to a game plan for this proof inside the following two-column solution, between the numbered lines.

Statements	Reasons
1) $\overline{MD} \cong \overline{VI}$	1) Given.
2) $\overline{DX} \cong \overline{CV}$	2) Given.

I expect you know what comes next, but for the sake of argument, pretend you don't. Statement 3 has to use one or both of the givens. To see how you can use the four segments from the givens, make up arbitrary lengths for the segments: say $\overline{MD}$ and $\overline{VI}$ both have a length of 5, and $\overline{DX}$ and $\overline{CV}$ are both 2. Obviously, that makes both $\overline{MX}$ and $\overline{CI}$ equal to 7, and that's called addition, of course. So now you've got line 3.

3) $\overline{MX} \cong \overline{CI}$	3) If two congruent segments are added to two other congruent segments, then the sums are congruent.

Now imagine that $\overline{XC}$ is 10. That would make both $\overline{MC}$ and $\overline{XI}$ equal to 17, and thus they're congruent. This is the three-segment version of segment addition, and that's a wrap.

4) $\overline{MC} \cong \overline{XI}$	4) If a segment is added to two congruent segments, then the sums are congruent.

Trivia answer: René Descartes, born in 1596.

Before looking at the next example, check out these two tips — they're huge! They can often make a tricky problem much easier and get you unstuck when you're stuck:

- **Use every given.** You have to use every given in a proof. So if you feel stuck while working on a proof, don't give up until you've asked, "Why did they give me this given?" for every one of the givens. If you then write down what follows from each given (even if you don't know how that will help you), you might see how to proceed. Don't forget: *every given is a built-in hint.*

- **Work backward.** Thinking about how a proof will end — what the last and second-to-last lines will look like — is often helpful. In some proofs, you may be able to work backward from the final statement to the second-to-last statement and then to the third-to-last statement and maybe even to the fourth-to-last. This makes the proof easier to finish because you no longer have to "see" all the way from the *given* to the *prove statement*. The proof has, in a sense, been shortened. Try this process when you get stuck in the middle of a proof; and sometimes it's a great thing to try when you begin to tackle a proof.

The following proof shows how you use angle addition:

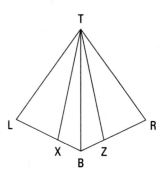

Given: $\overrightarrow{TB}$ bisects $\angle XTZ$

$\overrightarrow{TX}$ and $\overrightarrow{TZ}$ trisect $\angle LTR$

Prove: $\overrightarrow{TB}$ bisects $\angle LTR$

In this proof, I've added a partial game plan that deals with the part of the proof where people might get stuck.

Statements	Reasons
1) $\overrightarrow{TB}$ bisects $\angle XTZ$	1) Given. (Why would they tell you this? See statement 2.)
2) $\angle XTB \cong \angle ZTB$	2) If an angle is bisected, then it's divided into two congruent angles (definition of bisect).
3) $\overrightarrow{TX}$ and $\overrightarrow{TZ}$ trisect $\angle LTR$	3) Given. (And why would they tell you that?)
4) $\angle LTX \cong \angle RTZ$	4) If an angle is trisected, then it's divided into three congruent angles (definition of trisect).

(continued)

(continued)

Say you're stuck here. Try jumping to the end of the proof and working backward. You know that the final statement must be the *prove* conclusion, $\overline{TB}$ bisects $\angle LTR$. Now ask yourself what you'd need to know in order to draw that final conclusion. To conclude that a ray bisects an angle, you need to know that the ray cuts the angle into two equal angles. So the second-to-last statement must be $\angle LTB \cong \angle RTB$. And how do you deduce that? Well, with angle addition. The congruent angles from statements 2 and 4 add up to $\angle LTB$ and $\angle RTB$. That does it.

5) $\angle LTB \cong \angle RTB$	5) If two congruent angles are added to two other congruent angles, then the sums are congruent.
6) $\overrightarrow{TB}$ bisects $\angle LTR$	6) If a ray divides an angle into two congruent angles, then it bisects the angle (definition of bisect).

Subtraction theorems

Each of the following subtraction theorems corresponds to one of the addition theorems.

- ✔ **Segment subtraction (three total segments):** If a segment is subtracted from two congruent segments, then the differences are congruent.

- ✔ **Angle subtraction (three total angles):** If an angle is subtracted from two congruent angles, then the differences are congruent.

- ✔ **Segment subtraction (four total segments):** If two congruent segments are subtracted from two other congruent segments, then the differences are congruent.

- ✔ **Angle subtraction (four total angles):** If two congruent angles are subtracted from two other congruent angles, then the differences are congruent.

You use one of the *addition theorems* when you add *small* segments (or angles) and conclude that two *big* segments (or angles) are congruent. You use one of the *subtraction theorems* when you subtract segments (or angles) from *big* segments (or angles) to conclude that two *small* segments (or angles) are congruent.

Like Multiples and Like Divisions

The two theorems in this section are based on very simple ideas (multiplication and division), but they can be a bit tricky, so make sure to pay attention to the three tips.

Like Multiples: If two segments (or angles) are congruent, then their *like multiples* are congruent. For example, if you have two congruent angles, then three times one will equal three times the other.

Like Divisions: If two segments (or angles) are congruent, then their *like divisions* are congruent. If you have, say, two congruent segments, then $\frac{1}{4}$ of one equals $\frac{1}{4}$ of the other, or $\frac{1}{10}$ of one equals $\frac{1}{10}$ of the other, and so on.

Look at Figure 2-7. If $\angle BAC \cong \angle YXZ$ and both angles are bisected, then the Like Divisions Theorem tells you that $\angle 1 \cong \angle 3$ and that $\angle 2 \cong \angle 4$. And you could also use the theorem to deduce that $\angle 1 \cong \angle 4$ and that $\angle 2 \cong \angle 3$. But note that you *cannot* use the Like Divisions Theorem to conclude that $\angle 1 \cong \angle 2$ or $\angle 3 \cong \angle 4$. Those congruencies follow from the definition of bisect.

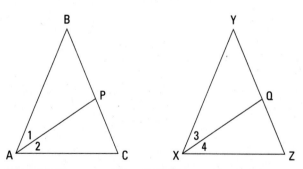

Figure 2-7: Congruent angles divided into congruent parts.

In a proof, you use the Like Multiples Theorem when you use congruent *small* segments (or angles) to conclude that two *big* segments (or angles) are congruent. You use the Like Divisions Theorem when you use congruent *big* things to conclude that two *small* things are congruent.

When you look at the givens in a proof and you see one of the terms *midpoint, bisect,* or *trisect* mentioned *twice,* then you'll probably use either the Like Multiples Theorem or the Like Divisions Theorem. But if the term is used only once, you'll likely use the definition of that term instead.

The following proof uses *Like Divisions*:

Given: $\overline{ND} \cong \overline{EL}$

O is the midpoint of $\overline{NE}$

A is the midpoint of $\overline{DL}$

Prove: $\overline{NO} \cong \overline{AL}$

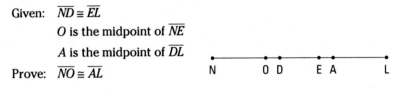

Here's a possible game plan: What can you do with the first given? If you can't figure that out right away, make up lengths for $\overline{ND}$, $\overline{EL}$, and $\overline{DE}$. Say that $\overline{ND}$ and $\overline{EL}$ are both 12 and that $\overline{DE}$ is 6. That would make both $\overline{NE}$ and $\overline{DL}$ 18 units long. Then, because both of these segments are bisected by their mid-points, $\overline{NO}$ and $\overline{AL}$ must both be 9. That's a wrap.

Statements	Reasons
1) $\overline{ND} \cong \overline{EL}$	1) Given.
2) $\overline{NE} \cong \overline{DL}$	2) If a segment is added to two congruent segments, then the sums are congruent.
3) O is the midpoint of $\overline{NE}$ A is the midpoint of $\overline{DL}$	3) Given.
4) $\overline{NO} \cong \overline{AL}$	4) If two segments are congruent ($\overline{NE}$ and $\overline{DL}$), then their like divisions are congruent (half of one equals half of the other).

The Like Divisions Theorem is particularly easy to get confused with the definitions of *midpoint, bisect,* and *trisect,* so remember this: Use the definition of *midpoint, bisect,* or *trisect* when you want to show that two parts of *one* bisected or trisected segment or angle are equal to each other. Use Like Divisions when *two* objects are bisected or trisected (like $\overline{NE}$ and $\overline{DL}$ in the preceding proof) and you want to show that a part of one ($\overline{NO}$) is equal to a part of the other ($\overline{AL}$).

Congruent Vertical Angles

Vertical angles are congruent: If two angles are vertical angles (angles on the opposite sides of an X), then they're congruent.

Here's an algebraic geometry problem that illustrates this simple concept: Determine the measure of the six angles in the following figure, given that $\angle 1 = (5x + 2y)°$, $\angle 2 = (-6x)°$, $\angle 4 = (2x + y)°$, and $\angle 5 = (y + 15)°$.

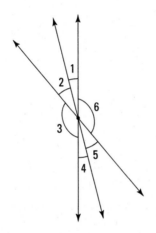

Vertical angles are congruent, so $\angle 1 \cong \angle 4$ and $\angle 2 \cong \angle 5$; and thus you can set their measures equal to each other:

$$\angle 1 \cong \angle 4 \qquad \angle 2 \cong \angle 5$$
$$5x + 2y = 2x + y \quad \text{and} \quad -6x = y + 15$$

Now you have a system of two equations and two unknowns. To solve the system, first solve each equation for y:

$$y = -3x \qquad y = -6x - 15$$

Next, because both equations are solved for y, you can set the two x-expressions equal to each other and solve for x:

$$-3x = -6x - 15$$
$$3x = -15$$
$$x = -5$$

To get y, plug in –5 for x in the first simplified equation:

$$y = -3x$$
$$y = -3(-5)$$
$$y = 15$$

Now plug in –5 and 15 to get four of the six angles:

$$\angle 4 \cong \angle 1 = 5x + 2y = 5(-5) + 2(15) = 5°$$
$$\angle 5 \cong \angle 2 = -6x = -6(-5) = 30°$$

To get angle 3, note that angles 1, 2, and 3 make a straight line, so they must sum to 180°:

$$\angle 1 + \angle 2 + \angle 3 = 180°$$
$$5° + 30° + \angle 3 = 180°$$
$$\angle 3 = 145°$$

Finally, $\angle 3$ and $\angle 6$ are congruent vertical angles, so $\angle 6$ must be 145° as well.

Transitivity and Substitution

The Transitive Property and the Substitution Property are two principles that you should understand right off the bat:

- ✔ **Transitive Property (for three segments or angles):** If two segments (or angles) are each congruent to a third segment (or angle), then they're congruent to each other. For example, if $\angle A \cong \angle B$ and $\angle B \cong \angle C$, then $\angle A \cong \angle C$.

- ✔ **Transitive Property (for four segments or angles):** If two segments (or angles) are congruent to congruent segments (or angles), then they're congruent to each other. For example, if $\overline{AB} \cong \overline{CD}$, $\overline{CD} \cong \overline{EF}$, and $\overline{EF} \cong \overline{GH}$, then $\overline{AB} \cong \overline{GH}$.

- ✔ **Substitution Property:** If two geometric objects (segments, angles, triangles, or whatever) are congruent and you have a statement involving one of them, you can pull the switcheroo and replace the one with the other. For example, if $\angle X \cong \angle Y$ and $\angle Y$ is supplementary to $\angle Z$, then $\angle X$ is supplementary to $\angle Z$.

To avoid getting the Transitive and Substitution Properties mixed up, follow the guidelines below, which are illustrated in the proofs that follow:

- ✔ Use the *Transitive Property* as the reason in a proof when the statement on the same line says things are congruent.

- ✔ Use the *Substitution Property* when the statement says something other than a congruence.

Check out this *TGIF* rectangle proof, which deals with angles:

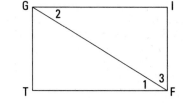

Given: ∠*TFI* is a right angle

∠1 ≅ ∠2

Prove: ∠2 is complementary to ∠3

Statements	Reasons
1) ∠*TFI* is a right angle	1) Given.
2) ∠1 is complementary to ∠3	2) If two angles form a right angle, then they're complementary (definition of complementary).
3) ∠1 is congruent to ∠2	3) Given.
4) ∠2 is complementary to ∠3	4) Substitution Property (statements 2 and 3; ∠2 replaces ∠1).

And for the final segment of the program, here's a related proof, *OSIM* (Oh Shoot, It's Monday):

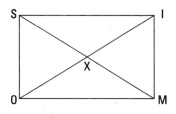

Given: *X* is the midpoint of $\overline{MS}$ and $\overline{OI}$

$\overline{SX} \cong \overline{IX}$

Prove: $\overline{MX} \cong \overline{OX}$

Statements	Reasons
1) X is the midpoint of $\overline{MS}$ and $\overline{OI}$	1) Given.
2) $\overline{SX} \cong \overline{MX}$ $\overline{IX} \cong \overline{OX}$	2) A midpoint divides a segment into two congruent segments.
3) $\overline{SX} \cong \overline{IX}$	3) Given.
4) $\overline{MX} \cong \overline{OX}$	4) Transitive Property (for four segments; statements 2 and 3).

Chapter 3

Tackling a Longer Proof

In This Chapter

▶ Making a game plan

▶ Starting at the start, working from the end, and meeting in the middle

▶ Making sure your logic holds

*C*hapter 2 started you off with short proofs and a dozen and a half basic theorems. Here, I go through a single, longer proof in great detail, carefully analyzing each step. Throughout the chapter, I walk you through the entire thought process that goes into solving a proof, reviewing and expanding on the half dozen or so proof strategies from Chapter 2. When you're working on a proof and you get stuck, this chapter is a good one to come back to for tips on how to get moving again. The proof I've created for this chapter isn't so terribly gnarly; it's just a bit longer than the ones in Chapter 2. Here it is:

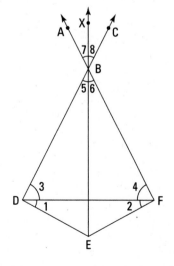

Given: $\overline{BD} \perp \overline{DE}$

$\overline{BF} \perp \overline{FE}$

$\angle 1 \cong \angle 2$

$\angle 5$ is complementary to $\angle 3$

$\angle 6$ is complementary to $\angle 4$

Prove: $\overrightarrow{BX}$ bisects $\angle ABC$

Making a Game Plan

A good way to begin any proof is to make a *game plan,* or rough outline, of how you'd do the proof. A game plan is your common sense argument of how you'd get from the given facts to the thing you're trying to prove. The nice thing about a game plan is that you don't need to worry about what theorems you'll need to use or how you'll write out the formal proof.

 When you're working through a game plan, it's sometimes a good idea to make up arbitrary numbers for the segments and angles in the givens and for unmentioned segments and angles. You should not, however, make up numbers for segments and angles that you're trying to show are congruent. This optional step makes the proof diagram more concrete and makes it easier for you to get a handle on how the proof works.

Here's one possible game plan for the proof we're working on: The givens provide you with two pairs of perpendicular segments; that gives you 90° for ∠BDE and ∠BFE. Then, say congruent angles ∠1 and ∠2 are both 30°. That would make ∠3 and ∠4 both equal to 60°. Next, because ∠3 and ∠5 are complementary, as are ∠4 and ∠6, ∠5 and ∠6 would both be 30°. Angles 5 and 8 are congruent vertical angles, as are ∠6 and ∠7, so ∠7 and ∠8 would also have to be 30° — and thus they're congruent. Finally, because ∠7 ≅ ∠8, ∠ABC is bisected. That does it.

Using All the Givens

Perhaps you don't follow the game plan in the previous section — or you get it but don't think you would've been able to come up with it on your own in one shot — and so you're staring at the proof and just don't know where to begin. My advice: Check all the givens in the proof and ask yourself *why* they'd tell you each given.

 Every given is a built-in hint. Look at the five givens in this proof. It's not immediately clear how the third, fourth, and fifth givens can help you, but what about the first two about the perpendicular segments? Why would they tell you this? What do perpendicular lines give you? Right angles, of course.

Okay, so you're on your way — you know the first two lines of the proof (see Figure 3-1).

Statements	Reasons
1) $\overline{BD} \perp \overline{DE}$ $\overline{BF} \perp \overline{FE}$	1) Given.
2) $\angle BDE$ is a right angle $\angle BFE$ is a right angle	2) If segments are perpendicular, then they form right angles.

Figure 3-1: The first two lines of the proof.

Using If-Then Logic

Moving from the givens to the final conclusion in a two-column proof is like knocking over a row of dominoes: Just as each domino knocks over the next domino, each proof statement leads to the next statement. The if-then sentence structure of each reason in a two-column proof shows you how each statement "knocks over" the next statement. In Figure 3-1, for example, you see the reason "*if* two segments are perpendicular, *then* they form a right angle." The perpendicular domino (statement 1) knocks over the right-angle domino (statement 2). This process continues throughout the whole proof.

Make sure that the if-then structure of your reasons is correct.

✔ The idea or ideas in the *if* clause of a reason must appear in the statement column somewhere *above* the line of that reason.

✔ The single idea in the *then* clause of a reason must be the same idea that's in the statement *directly across from* the reason.

Look back at Figure 3-1. Because statement 1 is the only statement above reason 2, it's the only place you can look for the ideas that go in the *if* clause of reason 2. So if you begin this proof by putting the two pairs of perpendicular segments in statement 1, then you have to use that information in reason 2, which must therefore begin "if segments are perpendicular, then . . ."

Now say you didn't know what to put in statement 2. The if-then structure of reason 2 helps you out. Because reason 2 begins "if two segments are perpendicular . . ." you'd ask yourself, "Well, what happens when two segments are perpendicular?" The answer, of course, is that right angles are formed. The right-angle idea must therefore go in the *then* clause of reason 2 and right across from it in statement 2.

Okay, now what? Well, think about reason 3. One way it could begin is with the right angles from statement 2. The *if* clause of reason 3 might be "if two angles are right angles . . ." Can you finish that? Of course: If two angles are right angles, then they're congruent. So that's it: You've got reason 3, and statement 3 must contain the idea from the *then* clause of reason 3, the congruence of right angles. Figure 3-2 shows you the proof so far.

Statements	Reasons
1) $\overline{BD} \perp \overline{DE}$ $\overline{BF} \perp \overline{FE}$	1) Given.
2) $\angle BDE$ is a right angle $\angle BFE$ is a right angle	2) If segments are perpendicular, then they form right angles.
3) $\angle BDE \cong \angle BFE$	3) If two angles are right angles, then they're congruent.

Figure 3-2: The first three lines of the proof.

When writing proofs, you need to spell out every little step as if you had to make the logic clear to a computer. For example, it may seem obvious that if you have two pairs of perpendicular segments, you've got congruent right angles, but this simple deduction takes three steps in a two-column proof. You have to go from perpendicular segments to right angles and then to congruent right angles — you can't jump straight to the congruent right angles. That's the way computers "think": A leads to B, B leads to C, and so on.

Chipping Away at the Problem

Face it: You're going to get stuck at one point or another while working on proofs. Here's a tip for getting unstuck.

Try something! When doing proofs, you need to be willing to experiment with ideas using trial and error. Doing proofs isn't as black and white as the math you've done before. You often can't know for sure what'll work. Just try something, and if it doesn't work, try something else. Sooner or later, the whole proof should fall into place.

So far in the proof in this chapter, you have the two congruent angles in statement 3, but you can't make more progress with that idea alone. So check out the givens again. Which of the three unused givens might build on statement 3? There's no way to know for sure, so you need to trust your instincts, pick a given, and try it.

The third given says ∠1 ≅ ∠2. That looks promising because angles 1 and 2 are part of the right angles from statement 3. You should ask yourself, "What would follow if ∠1 and ∠2 were, say, 35°?" You know the right angles are 90°, so if ∠1 and ∠2 were 35°, then ∠3 and ∠4 would both have to be 55° and thus, obviously, they'd be congruent. That's it. You're making progress. You can use that third given in statement 4 and then state that ∠3 ≅ ∠4 in statement 5.

Figure 3-3 shows the proof up to statement 5. The bubbles and arrows show you how the statements and reasons connect to each other. You can see that the *if* clause of each reason connects to a statement from above the reason and that the *then* clause connects to the statement on the same line as the reason. Because I haven't gone over reason 5 yet, it's not in the figure. See whether you can figure out reason 5 before reading the explanation that follows.

So, did you figure out reason 5? It's angle *subtraction* because ∠3 and ∠4 in statement 5 ended up being 55° angles (assuming ∠1 and ∠2 were 35°), and you get the answer of 55° by doing a subtraction problem, 90° − 35° = 55°. You're subtracting two angles from two other angles, so you use the four-angle version of angle subtraction (see Chapter 2). Reason 5 is, therefore, "If two congruent angles (∠1 and ∠2) are subtracted from two other congruent angles (the right angles), then the differences (∠3 and ∠4) are congruent."

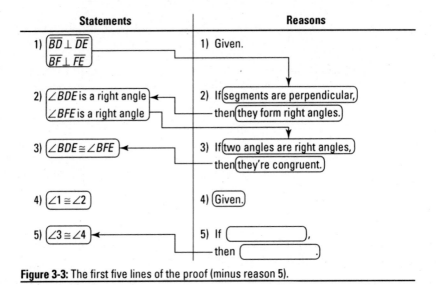

Figure 3-3: The first five lines of the proof (minus reason 5).

At this stage, you may feel a bit disconcerted if you don't know where these five lines are taking you or whether they're correct or not. Here's a tip.

If you're in the middle of solving a proof and can't see how to get to the end, remember that taking steps is a good thing. If you're able to deduce more and more facts and can begin filling in the statement column, you're very likely on the right path. Don't worry about the possibility that you're going the wrong way. (Although such detours do happen from time to time, don't sweat it. If you hit a dead end, just go back and try a different tack.) Don't feel like you've got to score a touchdown (that is, see how the whole proof fits together). Instead, be content with just making a first down (getting one more statement), then another first down, then another, and so on.

Working Backward

Assume that you're in the middle of a proof and you can't see how to get to the finish line from where you are now. No worries — just jump to the end of the proof and *work backward*.

Okay, so picking up where I left off on this chapter's proof: You've completed five lines of the proof, and you're up to $\angle 3 \cong \angle 4$. Where to now? Going forward from here might be a bit tricky, so work backward. You know that the final line of the proof has to be the *prove* statement: $\overrightarrow{BX}$ bisects $\angle ABC$. Now, if you think about what the final reason has to be or what the second-to-last statement should be, it shouldn't be too hard to see that you need to have two congruent angles to conclude that a larger angle is bisected. Figure 3-4 shows you what the end of the proof looks like.

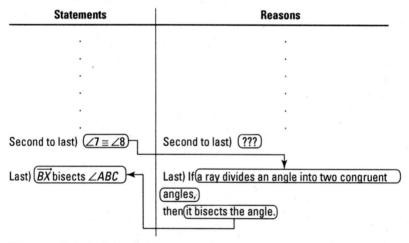

Figure 3-4: The proof's last two lines.

Try to continue going backward to the third-to-last statement, the fourth-to-last statement, and so on. (Working backward through a proof always involves some guesswork, but don't let that stop you.) Why might $\angle 7$ be congruent to $\angle 8$?

Well, you probably don't have to look too hard to spot the pair of congruent vertical angles ∠5 and ∠8 and the other pair, ∠6 and ∠7. Okay, so you want to show that ∠7 is congruent to ∠8, and you know that ∠6 equals ∠7 and ∠5 equals ∠8. So if you were to know that ∠5 and ∠6 are congruent, you'd be home free.

Now that you've worked backward a number of steps, here's the argument in the forward direction: The proof could end by stating in the fourth-to-last statement that ∠5 ≅ ∠6, then in the third-to-last that ∠5 ≅ ∠8 and ∠6 ≅ ∠7 (because vertical angles are congruent), and then in the second-to-last that ∠7 ≅ ∠8 by the Transitive Property (for four angles). Figure 3-5 shows how this all looks written out in the two-column format.

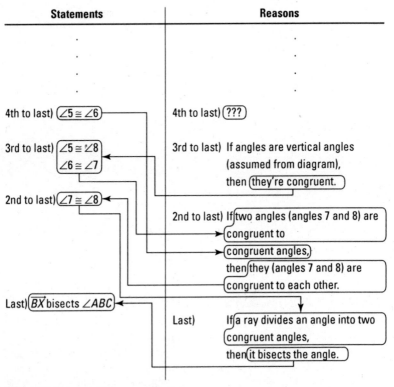

Statements	Reasons
. . .	. . .
4th to last) ∠5 ≅ ∠6	4th to last) ???
3rd to last) ∠5 ≅ ∠8, ∠6 ≅ ∠7	3rd to last) If angles are vertical angles (assumed from diagram), then they're congruent.
2nd to last) ∠7 ≅ ∠8	2nd to last) If two angles (angles 7 and 8) are congruent to congruent angles, then they (angles 7 and 8) are congruent to each other.
Last) $\overrightarrow{BX}$ bisects ∠ABC	Last) If a ray divides an angle into two congruent angles, then it bisects the angle.

Figure 3-5: The end of the proof (so far).

Filling In the Gaps

Working backward from the end of a proof is a great strategy. You can't always work as far backward as I did in this proof — sometimes you can only get to the second-to-last statement or maybe to the third-to-last. But even if you fill in only one or two statements (in addition to the automatic final statement), those additions can be very helpful. After making the additions, the proof is easier to finish because your new "final" destination (say the third-to-last statement) is fewer steps away from the beginning of the proof and is thus an easier goal to aim for.

Okay, let's wrap up this proof. All that remains is to bridge the gap between statement 5 ($\angle 3 \cong \angle 4$), and the fourth-to-last statement ($\angle 5 \cong \angle 6$). There are two givens you haven't used yet, so they must be the key to finishing the proof.

How can you use the givens about the two pairs of complementary angles? Try the plugging-in-numbers idea again. Use the same numbers we used before, and say that congruent angles $\angle 3$ and $\angle 4$ are each 55°. Angle 5 is complementary to $\angle 3$, so $\angle 5$ would have to be 35°. Angle 6 is complementary to $\angle 4$, so $\angle 6$ also ends up being 35°. That does it — $\angle 5$ and $\angle 6$ are congruent, and you've connected the loose ends. All that's left is to finish writing out the formal proof.

Writing Out the Finished Proof

Sound the trumpets! Here's the finished proof complete with the flow-of-logic bubbles (see Figure 3-6). (This time, I've put in only the arrows that connect to the *if* clause of each reason. You know that each reason's *then* clause must connect to the statement on the same line.) If you understand all the strategies and tips covered in this chapter and you can follow every step of this proof, you should be able to handle any proof they throw at you.

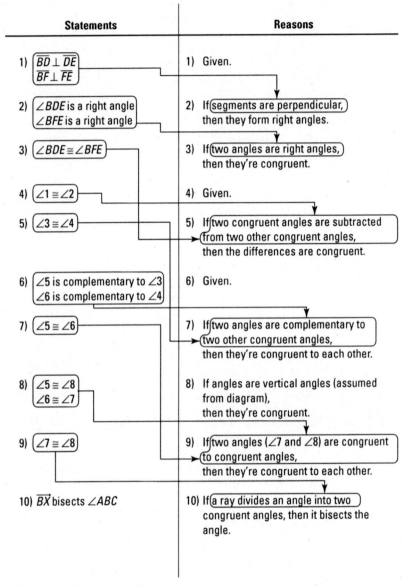

Statements	Reasons
1) $\overline{BD} \perp \overline{DE}$ $\overline{BF} \perp \overline{FE}$	1) Given.
2) ∠*BDE* is a right angle ∠*BFE* is a right angle	2) If segments are perpendicular, then they form right angles.
3) ∠*BDE* ≅ ∠*BFE*	3) If two angles are right angles, then they're congruent.
4) ∠1 ≅ ∠2	4) Given.
5) ∠3 ≅ ∠4	5) If two congruent angles are subtracted from two other congruent angles, then the differences are congruent.
6) ∠5 is complementary to ∠3 ∠6 is complementary to ∠4	6) Given.
7) ∠5 ≅ ∠6	7) If two angles are complementary to two other congruent angles, then they're congruent to each other.
8) ∠5 ≅ ∠8 ∠6 ≅ ∠7	8) If angles are vertical angles (assumed from diagram), then they're congruent.
9) ∠7 ≅ ∠8	9) If two angles (∠7 and ∠8) are congruent to congruent angles, then they're congruent to each other.
10) $\overrightarrow{BX}$ bisects ∠*ABC*	10) If a ray divides an angle into two congruent angles, then it bisects the angle.

Figure 3-6: The finished proof.

Chapter 4

Triangle Fundamentals

In This Chapter

▶ Classifying triangles by their angles

▶ Uncovering the triangle inequality principle

▶ Poring over the Pythagorean Theorem

*C*onsidering that it's the runt of the polygon family, the triangle sure does play a big role in geometry. Triangles are one of the most important components of geometry proofs. They also have a great number of interesting properties that you might not expect from the simplest possible polygon. Maybe Leonardo da Vinci (1452–1519) was on to something when he said, "Simplicity is the ultimate sophistication."

Taking In a Triangle's Sides

Triangles are classified according to the length of their sides or the measure of their angles. These classifications come in threes, just like the sides and angles themselves.

The following are triangle classifications based on sides:

✔ **Scalene:** A triangle with no congruent sides

✔ **Isosceles:** A triangle with at least two congruent sides

✔ **Equilateral:** A triangle with three congruent sides

Scalene triangles

In addition to having three unequal sides, scalene triangles have three unequal angles. The shortest side is across from the smallest angle, the medium side is across from the medium angle, and the longest side is across from the largest angle.

Isosceles triangles

An isosceles triangle has two (or three) equal sides and two (or three) equal angles. The equal sides are called *legs,* and the third side is the *base.* The two angles touching the base (which are congruent) are called *base angles.* The angle between the two legs is called the *vertex angle.* See Figure 4-1.

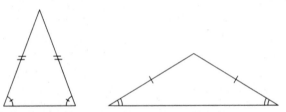

Figure 4-1: Two run-of-the-mill isosceles triangles.

Equilateral triangles

An equilateral triangle has three equal sides and three equal angles (which are each 60°). Its equal angles make it *equiangular* as well as equilateral. Note that an equilateral triangle is also isosceles.

Triangle Classification by Angles

✔ **Acute triangle:** A triangle with three acute angles.

✔ **Obtuse triangle:** A triangle with one obtuse angle. The other two angles are acute.

✔ **Right triangle:** A triangle with a single right angle and two acute angles. The *legs* of a right triangle are the sides touching the right angle, and the *hypotenuse* is the side across from the right angle.

The Triangle Inequality Principle

The triangle inequality principle: The sum of the lengths of any two sides of a triangle must be greater than the length of the third side. This principle comes up in a fair number of problems, so don't forget it! It's based on the simple fact that the shortest distance between two points is a straight line. Check out Figure 4-2 and the explanation that follows.

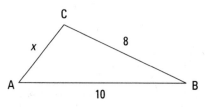

Figure 4-2: The triangle inequality principle lets you find the possible lengths of side $\overline{AC}$.

In $\triangle ABC$, what's the shortest route from A to B? Naturally, going straight across from A to B is shorter than taking a detour by traveling from A to C and then on to B. That's the triangle inequality principle in a nutshell.

In $\triangle ABC$, because you know that AB must be less than AC plus CB, $x + 8$ must be greater than 10; therefore,

$$x + 8 > 10$$
$$x > 2$$

But don't forget that the same principle applies to the path from A to C; thus, $8 + 10$ must be greater than x:

$$8 + 10 > x$$
$$18 > x$$

You can write both of these answers as a single inequality:

$$2 < x < 18$$

These are the possible lengths of side $\overline{AC}$. Note that the 2 and the 18 come from the difference and the sum of the other two sides. Figure 4-3 shows this range of lengths. Think of vertex B as a hinge. As the hinge opens more and more, the length of $\overline{AC}$ grows.

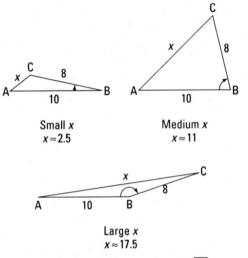

Small x
x ≈ 2.5

Medium x
x ≈ 11

Large x
x ≈ 17.5

Figure 4-3: Triangle ABC changes as side $\overline{AC}$ grows.

Sizing Up Triangle Area

I have a feeling you can guess what this section is about.

A triangle's altitude or height

Altitude (of a triangle): A segment from a vertex of a triangle to the opposite side (or to the extension of the opposite side if necessary) that's perpendicular to the opposite side. The opposite side is called the *base*.

Imagine that you have a cardboard triangle standing straight up on a table. The altitude of the triangle tells you exactly what you'd expect — the triangle's height *(h)* measured from its peak straight down to the table. This height goes down to the base of the triangle that's flat on the table. Figure 4-4 shows you an example of an altitude.

Every triangle has three altitudes, one for each side. Figure 4-5 shows the same triangle from Figure 4-4 standing up on a table in the other two possible positions: with $\overline{CB}$ as the base and with $\overline{BA}$ as the base.

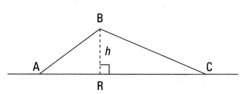

Figure 4-4: $\overline{BR}$ is one of the altitudes of $\triangle ABC$.

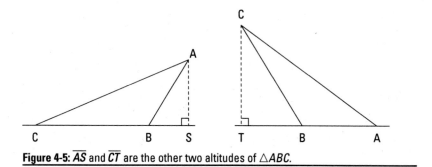

Figure 4-5: $\overline{AS}$ and $\overline{CT}$ are the other two altitudes of $\triangle ABC$.

You can use any side of a triangle as a base, regardless of whether that side is on the bottom. Figure 4-6 shows $\triangle ABC$ again with all three of its altitudes.

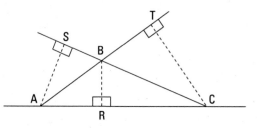

Figure 4-6: Triangle ABC with its three altitudes.

Here's the lowdown on the location of a triangle's altitudes:

- ✔ **Acute triangle:** All three altitudes are inside the triangle.
- ✔ **Right triangle:** One altitude is inside the triangle, and the other two altitudes are the legs of the triangle (remember this when figuring the area of a right triangle).
- ✔ **Obtuse triangle:** One altitude is inside the triangle, and two altitudes are outside the triangle.

Determining a triangle's area

Triangle area formula:

$$\text{Area}_\triangle = \frac{1}{2}\text{base} \cdot \text{height}$$

Assume for the sake of argument that you have trouble remembering this formula. Well, you won't forget it if you focus on why it's true — which brings me to one of the most important tips in this book.

Whenever possible, don't just memorize math concepts, formulas, and so on by rote. Try to understand *why* they're true. When you grasp the *why*s underlying the ideas, you remember them better and develop a deeper appreciation of the interconnections among mathematical ideas. That appreciation makes you a more successful math student.

So why does the area of a triangle equal $\frac{1}{2}$*base* · *height*?

Because the area of a rectangle is *base* · *height,* and a triangle is half of a rectangle. Check out Figure 4-7, which shows two triangles inscribed in rectangles *HALF* and *PINT.*

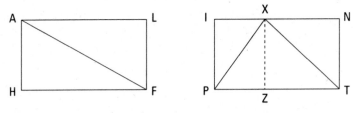

Figure 4-7: A triangle takes up half the area of a rectangle.

It should be really obvious that △*HAF* has half the area of rectangle *HALF.* And it shouldn't exactly give you a brain hemorrhage to see that △*PXT* also has half the area of the rectangle around it. (Triangle *PXZ* is half of rectangle *PIXZ,* and △*ZXT* is half of rectangle *ZXNT.*) Because every possible triangle fits in some rectangle just like △*PXT* fits in rectangle *PINT* (you just have to put the triangle's longest side on the bottom), every triangle is half a rectangle.

Now for a problem: What's the length of altitude $\overline{XT}$ in △*WXR* in Figure 4-8?

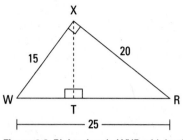

Figure 4-8: Right triangle WXR with its three altitudes.

The trick here is to note that because $\triangle WXR$ is a right triangle, legs $\overline{WX}$ and $\overline{RX}$ are also altitudes. So you can use either one as the altitude, and then the other leg automatically becomes the base. Plug their lengths into the formula to determine the triangle's area:

$$\text{Area}_{\triangle WXR} = \frac{1}{2}\text{base} \cdot \text{height}$$
$$= \frac{1}{2}(RX)(WX)$$
$$= \frac{1}{2}(20)(15)$$
$$= 150$$

Now you can use the area formula again, using this area of 150, base $\overline{WR}$, and altitude $\overline{XT}$:

$$\text{Area}_{\triangle WXR} = \frac{1}{2}\text{base} \cdot \text{height}$$
$$150 = \frac{1}{2}(WR)(XT)$$
$$150 = \frac{1}{2}(25)(XT)$$
$$12 = XT$$

Regarding Right Triangles

In the mathematical universe of all possible triangles, right triangles are extremely rare. But in the *real* world, right triangles are extremely common. Right angles are everywhere: the corners of almost every wall, floor, ceiling, door, and window; the corners of every book, table, and box; the intersection of most streets; and so on. And everywhere you see a right angle, you potentially have a right triangle. Right triangles abound in navigation, carpentry, and architecture — even the builders of the Great Pyramids in Egypt used right-triangle mathematics. The next section shows you the elegant relationship among the three sides of a right triangle.

The Pythagorean Theorem

The Pythagorean Theorem:

$$a^2 + b^2 = c^2$$

Here, a and b are the lengths of the legs and c is the length of the hypotenuse. The *legs* are the two short sides that touch the right angle, and the *hypotenuse* (the longest side) is opposite the right angle. Figure 4-9 shows how the Pythagorean Theorem works for a right triangle with legs of 3 and 4 and a hypotenuse of 5.

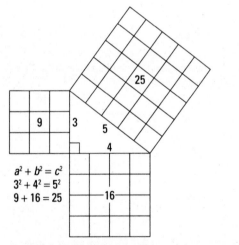

Figure 4-9: The Pythagorean Theorem is as easy as 9 + 16 = 25.

Here's a multistage problem in which you have to use the Pythagorean Theorem more than once: In Figure 4-10, find x and the area of hexagon *ABCDEF.*

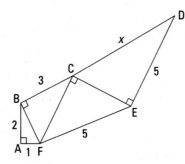

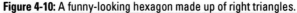

Figure 4-10: A funny-looking hexagon made up of right triangles.

ABCDEF is made up of four connected right triangles. To get *x*, you set up a chain reaction in which you solve for the unknown side of one triangle and then use that answer to find the unknown side of the next triangle, and so on. You already know the lengths of two sides of △*BAF*, so start there to find *BF*:

$$(BF)^2 = (AF)^2 + (AB)^2$$
$$(BF)^2 = 1^2 + 2^2$$
$$(BF)^2 = 5$$
$$BF = \sqrt{5}$$

Now that you have *BF*, you know two of the sides of △*CBF*. Use the Pythagorean Theorem to find *CF*:

$$(CF)^2 = (BF)^2 + (BC)^2$$
$$(CF)^2 = \sqrt{5}^2 + 3^2$$
$$(CF)^2 = 5 + 9$$
$$CF = \sqrt{14}$$

With *CF* filled in, you can find the short leg of △*ECF*:

$$(CE)^2 + (CF)^2 = (FE)^2$$
$$(CE)^2 + \sqrt{14}^2 = 5^2$$
$$(CE)^2 + 14 = 25$$
$$(CE)^2 = 11$$
$$CE = \sqrt{11}$$

And now that you know *CE*, you can solve for *x*:

$$x^2 = (CE)^2 + (ED)^2$$
$$x^2 = \sqrt{11}^2 + 5^2$$
$$x^2 = 36$$
$$x = 6$$

Okay, on to the second half of the problem. To get the area of *ABCDEF*, just add up the areas of the four right triangles. The area of a triangle is $\frac{1}{2} \cdot$ **base** $\cdot$ **height**. For a right triangle, you can use the two legs for the base and the height. Solving for *x* has already given you the lengths of all the sides of the triangles, so just plug the numbers into the area formula:

$$\text{Area}_{\triangle BAF} = \frac{1}{2} \cdot 1 \cdot 2 \qquad \text{Area}_{\triangle CBF} = \frac{1}{2} \cdot \sqrt{5} \cdot 3$$
$$= 1 \qquad\qquad\qquad = 1.5\sqrt{5}$$
$$\text{Area}_{\triangle ECF} = \frac{1}{2} \cdot \sqrt{11} \cdot \sqrt{14} \quad \text{Area}_{\triangle DCE} = \frac{1}{2} \cdot \sqrt{11} \cdot 5$$
$$= 0.5\sqrt{154} \qquad\qquad = 2.5\sqrt{11}$$

Thus, the area of hexagon *ABCDEF* is $1 + 1.5\sqrt{5} + 0.5\sqrt{154} + 2.5\sqrt{11}$, or about 18.9 units^2.

Pythagorean Triple Triangles

If you use any old numbers for two sides of a right triangle, the Pythagorean Theorem almost always gives you the square root of something for the third side. For example, a right triangle with legs of 5 and 6 has a hypotenuse of $\sqrt{61}$; if the legs are 3 and 8, the hypotenuse is $\sqrt{73}$; and if one of the legs is 6 and the hypotenuse is 9, the other leg works out to $\sqrt{81-36}$, which is $\sqrt{45}$, or $3\sqrt{5}$.

A *Pythagorean triple triangle* is a right triangle with sides whose lengths are all whole numbers, such as 3, 4, and 5 or 5, 12, and 13. People like to use these triangles in problems because they don't contain those pesky square roots.

The Fab Four triangles

The first four Pythagorean triple triangles are the favorites of geometry problem makers. These triangles, especially the first and second on the list, pop up all over in geometry books.

Here are the first four Pythagorean triple triangles:

- ✔ The 3-4-5 triangle
- ✔ The 5-12-13 triangle
- ✔ The 7-24-25 triangle
- ✔ The 8-15-17 triangle

Families of Pythagorean triple triangles

Each irreducible Pythagorean triple triangle such as the 5-12-13 triangle is the matriarch of a family with an infinite number of children. The 3 : 4 : 5 *family* (note the colons), for example, consists of the 3-4-5 triangle and all her offspring. Offspring are created by blowing up or shrinking the 3-4-5 triangle: They include the $\frac{3}{100} - \frac{4}{100} - \frac{5}{100}$ triangle, the 6-8-10 triangle, the 21-28-35 triangle (3-4-5 times 7), and their eccentric siblings such as the $3\sqrt{11} - 4\sqrt{11} - 5\sqrt{11}$ triangle and the 3π-4π -5π triangle. Within any of the triangle families (like the 3 : 4 : 5 family), all the triangles have the same shape.

When you know two of the three sides of a right triangle, you can, of course, compute the third side with the Pythagorean Theorem. But if the triangle happens to be a member of one of the Fab Four families, you can use a shortcut. All you need to do is figure out the blow-up or shrink factor that converts the main Fab Four triangle into the given triangle and use that factor to compute the third side of the given triangle.

No-brainer cases

You can often just see that you have one of the Fab Four families and figure out the blow-up or shrink factor in your head. Check out Figure 4-11.

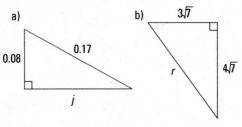

Figure 4-11: Two triangles from famous families.

In Figure 4-11a, the digits 8 and 17 in the 0.08 and 0.17 should give you a big hint that this triangle is a member of the $8 : 15 : 17$ family. Because 8 divided by 100 is 0.08 and 17 divided by 100 is 0.17, this triangle is an 8-15-17 triangle shrunk down 100 times. Side j is thus 15 divided by 100, or 0.15. This shortcut is way easier than using the Pythagorean Theorem.

Likewise, the digits 3 and 4 should make it a dead giveaway that the triangle in Figure 4-11b is a member of the $3 : 4 : 5$ family. Because $3\sqrt{7}$ is $\sqrt{7}$ times 3 and $4\sqrt{7}$ is $\sqrt{7}$ times 4, you can see that this triangle is a 3-4-5 triangle blown up by a factor of $\sqrt{7}$. Thus, side r is simply $\sqrt{7}$ times 5, or $5\sqrt{7}$.

Make sure the sides of the given triangle match up correctly with the sides of the Fab Four triangle family you're using. In a $3 : 4 : 5$ triangle, for example, the legs must be the 3 and the 4, and the hypotenuse must be the 5. So a triangle with legs of 30 and 50 (despite the 3 and the 5) is not in the $3 : 4 : 5$ family because the 50 (the 5) is one of the legs instead of the hypotenuse.

The step-by-step triple triangle method

If you can't immediately see what Fab Four family a triangle belongs to, you can always use the following step-by-step method to pick the family and find the missing side. Don't be put off by the length of the method; it's easier to do than to explain. Check out the triangle in Figure 4-12.

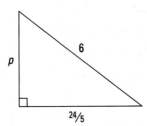

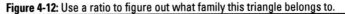

Figure 4-12: Use a ratio to figure out what family this triangle belongs to.

1. **Take the two known sides and make a ratio in fraction form of the smaller to the larger side.**

   Take the $\frac{24}{5}$ and the 6 and make the ratio of $\dfrac{24/5}{6}$.

2. **Reduce this ratio to whole numbers in lowest terms.**

   If you multiply the top and bottom of $\dfrac{24/5}{6}$ by 5, you get $\frac{24}{30}$; that reduces to $\frac{4}{5}$. (With many calculators, this is a snap because they have a function that reduces fractions to lowest terms.)

3. **Look at the fraction from Step 2 to spot the particular triangle family.**

   The numbers 4 and 5 are part of the 3-4-5 triangle, so you're dealing with the 3 : 4 : 5 family.

4. **Divide the length of a side from the given triangle by the corresponding number from the family ratio to get your multiplier (which tells you how much the basic triangle has been blown-up or shrunk).**

   Use the length of the hypotenuse from the given triangle (because working with a whole number is easier) and divide it by the 5 from the 3 : 4 : 5 ratio. You should get $\frac{6}{5}$ or 1.2 for your multiplier.

5. Multiply the third family number (the number you *don't* see in the reduced fraction in Step 2) by the result from Step 4 to find the missing side of your triangle.

Three times $\frac{6}{5}$ is $\frac{18}{5}$. That's the length of side p; and that's a wrap.

You may be wondering why you should go through all this trouble when you could just use the Pythagorean Theorem. Good point. The Pythagorean Theorem is easier for some triangles (especially if you're allowed to use your calculator). But — take my word for it — this triple triangle technique can come in handy. Take your pick.

Two Special Right Triangles

Make sure you know the two right triangles in this section: the 45°- 45°- 90° triangle and the 30°- 60°- 90° triangle. They come up in many, many geometry problems, not to mention their frequent appearance in trigonometry, precalculus, and calculus. Despite the pesky irrational (square-root) lengths they have for some of their sides, they're both more basic and more important than the Pythagorean triple triangles in the previous section. They're more basic because they're the progeny of the square and equilateral triangle, and they're more important because their angles are nice fractions of a right angle.

The 45°- 45°- 90° triangle

The 45°- 45°- 90° triangle (or isosceles right triangle): A triangle with angles of 45°, 45°, and 90° and sides in the ratio of $1:1:\sqrt{2}$. Note that it's the shape of half a square, cut along the square's diagonal, and that it's also an isosceles triangle. See Figure 4-13.

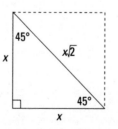

Figure 4-13: The 45°- 45°- 90° triangle.

Try a couple of problems. Find the lengths of the unknown sides in triangles BAT and BOY shown in Figure 4-14.

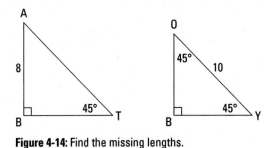

Figure 4-14: Find the missing lengths.

You can solve 45°- 45°- 90° triangle problems in two ways: the formal book method and the street-smart method. Try 'em both and take your pick. The formal method uses the ratio of the sides from Figure 4-13.

leg : leg : hypotenuse

$$x \ : \ x \ : x\sqrt{2}$$

For $\triangle BAT$, because one of the legs is 8, the x in the ratio is 8. Plugging 8 into the three x's gives you

leg : leg : hypotenuse
8 : 8 : $8\sqrt{2}$

And for $\triangle BOY$, the hypotenuse is 10, so you set the $x\sqrt{2}$ from the ratio equal to 10 and solve for x:

$$x\sqrt{2} = 10$$

$$x = \frac{10}{\sqrt{2}} = \frac{10}{\sqrt{2}} \cdot \frac{\sqrt{2}}{\sqrt{2}} = \frac{10\sqrt{2}}{2} = 5\sqrt{2}$$

That does it:

leg : leg : hypotenuse
$5\sqrt{2}$: $5\sqrt{2}$: 10

Now for *the street-smart method* (this uses the same math as the formal method, but it involves fewer steps): Remember the 45°- 45°- 90° triangle as the "$\sqrt{2}$ triangle." Using that tidbit, do one of the following:

- ✔ If you know a leg and want to compute the hypotenuse (a *longer* thing), you *multiply* by $\sqrt{2}$. In Figure 4-15, one of the legs in $\triangle BAT$ is 8, so you *multiply* that by $\sqrt{2}$ to get the *longer* hypotenuse — $8\sqrt{2}$.

- ✔ If you know the hypotenuse and want to compute the length of a leg (a *shorter* thing), you *divide* by $\sqrt{2}$. In Figure 8-8, the hypotenuse in $\triangle BOY$ is 10, so you *divide* that by $\sqrt{2}$ to get the *shorter* legs; they're each $\dfrac{10}{\sqrt{2}}$ or $5\sqrt{2}$.

The 30°- 60°- 90° triangle

The 30°- 60°- 90° triangle: A triangle with angles of 30°, 60°, and 90° and sides in the ratio of $1 : \sqrt{3} : 2$. Note that it's the shape of half an equilateral triangle, cut straight down the middle along its altitude. Check out Figure 4-15.

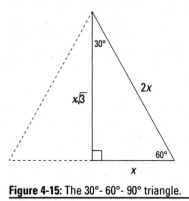

Figure 4-15: The 30°- 60°- 90° triangle.

Here are a couple of problems. Find the lengths of the unknown sides in $\triangle UMP$ and $\triangle IRE$ in Figure 4-16.

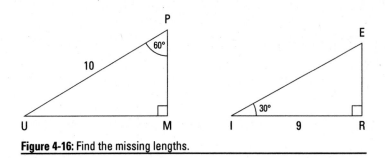

Figure 4-16: Find the missing lengths.

You can solve 30°- 60°- 90° triangles with the textbook method or the street-smart method. The textbook method begins with the ratio of the sides from Figure 4-15:

short leg : long leg : hypotenuse

x : $x\sqrt{3}$: $2x$

In △*UMP*, the hypotenuse is 10, so you set $2x$ equal to 10 and solve for x, getting $x = 5$. Now just plug 5 in for the x's, and you have △*UMP:*

short leg : long leg : hypotenuse

5 : $5\sqrt{3}$: 10

In △*IRE*, the long leg is 9, so set $x\sqrt{3}$ equal to 9 and solve:

$$x\sqrt{3} = 9$$

$$x = \frac{9}{\sqrt{3}} = \frac{9}{\sqrt{3}} \cdot \frac{\sqrt{3}}{\sqrt{3}} = \frac{9\sqrt{3}}{3} = 3\sqrt{3}$$

Plug in the value of x, and you're done:

short leg : long leg : hypotenuse

$\left(3\sqrt{3}\right)$: $\left(3\sqrt{3}\right)\sqrt{3}$: $2\left(3\sqrt{3}\right)$

$3\sqrt{3}$: 9 : $6\sqrt{3}$

Here's the *street-smart method*. Think of the 30°-60°-90° triangle as the "√3 triangle." Using that fact, do the following:

- ✔ The relationship between the short leg and the hypotenuse is a no-brainer: The hypotenuse is twice as long as the short leg. So if you know one of them, you can get the other in your head. The √3 method mainly concerns the connection between the short and long legs.

- ✔ If you know the short leg and want to compute the long leg (a *longer* thing), you *multiply* by √3. If you know the long leg and want to compute the short leg (a *shorter* thing), you *divide* by √3.

Try the street-smart method with the triangles in Figure 4-16. The hypotenuse in △*UMP* is 10, so first you cut that in half to get the length of the short leg, which is thus 5. Then to get the *longer* leg, you *multiply* that by √3, which gives you 5√3. In △ *IRE,* the long leg is 9, so to get the *shorter* leg, you *divide* that by √3, which gives you $\frac{9}{\sqrt{3}}$, or 3√3. The hypotenuse is twice that, 6√3.

With the 30°-60°-90° triangle (and also with the 45°-45°-90° triangle), there will almost always be one or two sides whose lengths contain a square root symbol (in unusual cases, all three sides could contain a radical symbol). But it's impossible to have no square roots — which brings me to the following warning.

Because at least one side of a 30°-60°-90° triangle must contain a square root, a 30°-60°-90° triangle cannot belong to any of the Pythagorean triple triangle families. So don't make the mistake of thinking that a 30°-60°-90° triangle is in, say, the 8 : 15 : 17 family or that any triangle that is in one of the Pythagorean triple triangle families is also a 30°-60°-90° triangle. There's no overlap between the 30°-60°-90° triangle (or the 45°-45°-90° triangle) and any of the Pythagorean triple triangles and their families.

Chapter 5

Congruent Triangle Proofs

*Y*ou've arrived at high school geometry's main event: triangle proofs. The proofs in Chapters 2 and 3 are complete proofs that show you how proofs work, and they illustrate many of the most important proof strategies. But they're sort of just warm-up or preliminary proofs that lay the groundwork for the real, full-fledged triangle proofs you see in this chapter.

Proving Triangles Congruent

Before learning how to prove that triangles are congruent, you've got to know what congruent triangles are, right? Here you go . . . .

Congruent triangles: Triangles in which all pairs of corresponding sides and angles are congruent.

Maybe the best way to think about what it means for two triangles (or any other shapes) to be congruent is that you could move them around (by shifting, rotating, and/or flipping them) so that they'd stack perfectly on top of one another. You indicate that triangles are congruent with a statement such as $\triangle ABC \cong \triangle XYZ$, which means that vertex *A* (the first letter) corresponds with and would stack on vertex *X* (the first letter), *B* would stack on *Y*, and *C* would stack on *Z*. Side $\overline{AB}$ would stack on side $\overline{XY}$, $\angle B$ would stack on $\angle Y$, and so on.

So now, on to the methods for proving triangles congruent. There are five ways: SSS, SAS, ASA, AAS, and HLR.

SSS: The side-side-side method

SSS (Side-Side-Side): If the three sides of one triangle are congruent to the three sides of another triangle, then the triangles are congruent. Figure 5-1 illustrates this idea.

If 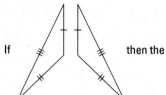 then the triangles are congruent.

Figure 5-1: Triangles with congruent sides are congruent.

You can use SSS in the following "*TRIANGLE*" proof:

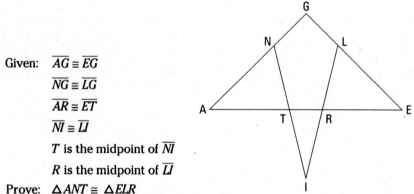

Given: $\overline{AG} \cong \overline{EG}$

$\overline{NG} \cong \overline{LG}$

$\overline{AR} \cong \overline{ET}$

$\overline{NI} \cong \overline{LI}$

T is the midpoint of $\overline{NI}$

R is the midpoint of $\overline{LI}$

Prove: $\triangle ANT \cong \triangle ELR$

First, come up with a game plan. Here's how that might work.

You know you've got to prove the triangles congruent, so your first question should be "Can you show that the three pairs of corresponding sides are congruent?" Sure, you can do that:

✔ Subtract $\overline{NG}$ and $\overline{LG}$ from $\overline{AG}$ and $\overline{EG}$ to get the first pair of congruent sides, $\overline{AN}$ and $\overline{EL}$.

> ✔ Subtract $\overline{TR}$ from $\overline{AR}$ and $\overline{ET}$ to get the second pair of congruent sides, $\overline{AT}$ and $\overline{ER}$.
>
> ✔ Cut congruent segments $\overline{NI}$ and $\overline{LI}$ in half to get the third pair, $\overline{NT}$ and $\overline{LR}$. That's it.

To make the game plan more tangible, you may want to make up lengths for the various segments. For instance, say $\overline{AG}$ and $\overline{EG}$ are 9, $\overline{NG}$ and $\overline{LG}$ are 3, $\overline{AR}$ and $\overline{ET}$ are 8, $\overline{TR}$ is 3, and $\overline{NI}$ and $\overline{LI}$ are 8. When you do the math, you see that $\triangle ANT$ and $\triangle ELR$ both end up with sides of 4, 5, and 6, which means, of course, that they're congruent.

Here's how the formal proof shapes up:

Statements	Reasons
1) $\overline{AG} \cong \overline{EG}$ $\overline{NG} \cong \overline{LG}$	1) Given.
2) $\overline{AN} \cong \overline{EL}$	2) If two congruent segments are subtracted from two other congruent segments, then the differences are congruent.
3) $\overline{AR} \cong \overline{ET}$	3) Given.
4) $\overline{AT} \cong \overline{ER}$	4) If a segment is subtracted from two congruent segments, then the differences are congruent.
5) $\overline{NI} \cong \overline{LI}$ T is the midpoint of $\overline{NI}$ R is the midpoint of $\overline{LI}$	5) Given.
6) $\overline{NT} \cong \overline{LR}$	6) If segments are congruent, then their Like Divisions are congruent (half of one equals half of the other — see Chapter 2).
7) $\triangle ANT \cong \triangle ELR$	7) SSS (2, 4, 6).

Note: After SSS in the final step, I indicate the three lines from the statement column where I've shown the three pairs of sides to be congruent. You don't have to do this, but it's a good idea. It can help you avoid some careless mistakes. Remember that each of the three lines you list must show a *congruence* of segments (or angles, if you're using one of the other approaches to proving triangles congruent).

SAS: side-angle-side

SAS (Side-Angle-Side): If two sides and the included angle of one triangle are congruent to two sides and the included angle of another triangle, then the triangles are congruent. (The *included angle* is the angle formed by the two sides.) Figure 5-2 illustrates this method.

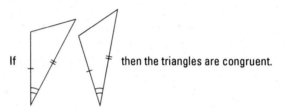

If ... then the triangles are congruent.

Figure 5-2: Two sides and the angle between them make these triangles congruent.

Check out the SAS postulate in action:

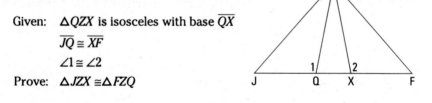

Given: △*QZX* is isosceles with base $\overline{QX}$

$\overline{JQ} \cong \overline{XF}$

∠1 ≅ ∠2

Prove: △*JZX* ≅ △*FZQ*

When overlapping triangles muddy your understanding of a proof diagram, try redrawing the diagram with the triangles separated. Doing so can give you a clearer idea of how the triangles' sides and angles relate to each other. Focusing on your new diagram may make it easier to figure out what you need to prove the triangles congruent. However, you still need to use the original diagram to understand some parts of the proof, so use the second diagram as a sort of aid to get a better handle on the original diagram.

Figure 5-3 shows you what this proof diagram looks like with the triangles separated.

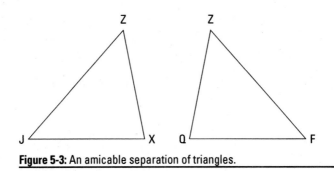

Figure 5-3: An amicable separation of triangles.

Looking at Figure 5-3, you can easily see that the triangles are congruent (they're mirror images of each other). You also see that, for example, side $\overline{ZX}$ corresponds to side $\overline{ZQ}$ and that $\angle X$ corresponds to $\angle Q$.

So using both diagrams, here's a possible game plan:

✔ **Determine which congruent triangle postulate is likely to be the ticket for proving the triangles congruent.** You know you have to prove the triangles congruent, and one of the givens is about angles, so SAS looks like a better candidate than SSS for the final reason. (You don't have to figure this out now, but it's not a bad idea to at least have a guess about the final reason.)

✔ **Look at the givens and think about what they tell you about the triangles.** Triangle QZX is isosceles, so that tells you $\overline{ZQ} \cong \overline{ZX}$. Look at these sides in both figures. Put tick marks on $\overline{ZQ}$ and $\overline{ZX}$ in Figure 5-3 to show that you know they're congruent. Now consider why they'd tell you the next given, $\overline{JQ} \cong \overline{XF}$. Well, what if they were both 6 and $\overline{QX}$ were 2? $\overline{JX}$ and $\overline{QF}$ would both be 8, so you have a second pair of congruent sides. Put tick marks on Figure 5-3 to show this congruence.

✔ **Find the pair of congruent angles.** Look at Figure 5-3 again. If you can show that $\angle X$ is congruent to $\angle Q$, you'll have SAS. Do you see where $\angle X$ and $\angle Q$ fit into the original diagram? Note that they're the supplements of $\angle 1$ and $\angle 2$. That does it. Angles 1 and 2 are congruent, so their supplements are congruent as well. (If you fill in numbers, you can see that if $\angle 1$ and $\angle 2$ are both 100°, $\angle Q$ and $\angle X$ would both be 80°.)

Here's the formal proof:

Statements	Reasons
1) △ *QZX* is isosceles with base $\overline{QX}$	1) Given.
2) $\overline{ZX} \cong \overline{ZQ}$	2) Definition of isosceles triangle.
3) $\overline{JQ} \cong \overline{XF}$	3) Given.
4) $\overline{JX} \cong \overline{FQ}$	4) If a segment is added to two congruent segments, then the sums are congruent.
5) ∠1 ≅ ∠2	5) Given.
6) ∠*ZXJ* ≅ ∠*ZQF*	6) If two angles are supplementary to two other congruent angles, then they're congruent.
7) △ *JZX* ≅ △ *FZQ*	7) SAS (2, 6, 4).

ASA: The angle-side-angle tack

ASA (Angle-Side-Angle): If two angles and the included side of one triangle are congruent to two angles and the included side of another triangle, then the triangles are congruent. See Figure 5-4.

If   then the triangles are congruent.

Figure 5-4: Two angles and their shared side make these triangles congruent.

AAS: angle-angle-side

AAS (Angle-Angle-Side): If two angles and a nonincluded side of one triangle are congruent to the corresponding parts of another triangle, then the triangles are congruent. Figure 5-5 shows you how AAS works.

If 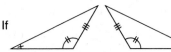 then the triangles are congruent.

Figure 5-5: Two congruent angles and a side not between them make these triangles congruent.

Like ASA, to use AAS, you need two pairs of congruent angles and one pair of congruent sides to prove two triangles congruent. But for AAS, the two angles and one side in each triangle must go in the order angle-angle-side (going around the triangle either clockwise or counterclockwise).

ASS and SSA don't prove anything, so don't try using ASS (or its backward twin, SSA) to prove triangles congruent. You can use SSS, SAS, ASA, and AAS (or SAA, the backward twin of AAS) to prove triangles congruent, but not ASS. In short, every three-letter combination of *A*'s and *S*'s proves something unless it spells *ass* or is *ass* backward. (You work with AAA in Chapter 8, but it shows that triangles are similar, not congruent.)

Last but not least: HLR

HLR (Hypotenuse-Leg-Right angle): If the hypotenuse and a leg of one right triangle are congruent to the hypotenuse and a leg of another right triangle, then the triangles are congruent. HLR is different than the other four ways of proving triangles congruent because it works only for right triangles.

In other books, HLR is usually called HL. Rebel that I am, I'm boldly renaming it HLR because its three letters emphasize that — as with SSS, SAS, ASA, and AAS — before you can use it in a proof, you need to have three things in the statement column (congruent hypotenuses, congruent legs, and right angles).

Taking the Next Step with CPCTC

In the preceding section, the relatively short proofs end with showing that two triangles are congruent. But in more advanced proofs, showing triangles congruent is just a stepping stone for going on to prove other things. In this section, you take proofs a step further.

Proving triangles congruent is often the focal point of a proof, so always check the proof diagram for *all* pairs of triangles that look like they're the same shape and size. If you find any, you'll very likely have to prove one (or more) of the pairs of triangles congruent.

Defining CPCTC

CPCTC: An acronym for *corresponding parts of congruent triangles are congruent.* This idea sort of has the feel of a theorem, but it's really just the definition of congruent triangles.

Because congruent triangles have six pairs of congruent parts (three pairs of segments and three pairs of angles) and you need three of the pairs for SSS, SAS, ASA, AAS, or HLR, there will always be three remaining pairs that you didn't use. The purpose of CPCTC is to show one or more of these remaining pairs congruent.

CPCTC is very easy to use. After you show that two triangles are congruent, you can state that two of their sides or angles are congruent on the next line of the proof, using CPCTC as the justification for that statement. This group of two consecutive lines makes up the core or heart of many proofs.

Say you're in the middle of some proof (shown in Figure 5-6), and by line 6, you're able to show with ASA that $\triangle PQR$ is congruent to $\triangle XYZ$. The tick marks in the diagram show the pair of congruent sides and the two pairs of congruent angles that were used for ASA. Now that you know that the triangles are congruent, you can state on line 7 that $\overline{QR} \cong \overline{YZ}$ and use CPCTC for the reason. (You could also use CPCTC to justify that $\overline{PR} \cong \overline{XZ}$ or that $\angle QRP \cong \angle YZX$.)

Tackling a CPCTC proof

Check out CPCTC in action in the next proof. But before I get there, here's a property you need to do the problem. It's an incredibly simple concept that comes up in many proofs.

The Reflexive Property: Any segment or angle is congruent to itself. (Who would've thought?)

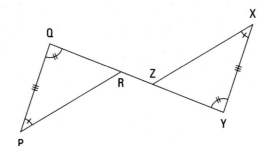

Statements	Reasons
...	...
...	...
...	...
6) $\triangle PQR \cong \triangle XYZ$	6) ASA.
7) $\overline{QR} \cong \overline{YZ}$	7) CPCTC.
...	...
...	...
...	...

Figure 5-6: A critical pair of proof lines: Congruent triangles and CPCTC.

Whenever you see two triangles that share a side or an angle, that side or angle belongs to both triangles. With the Reflexive Property, the shared side or angle becomes a pair of congruent sides or angles that you can use as one of the three pairs of congruent things that you need to prove the triangles congruent.

Here's your CPCTC proof:

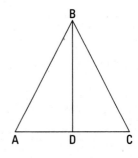

Given: $\overline{BD}$ is a median and
and altitude of $\triangle ABC$

Prove: $\overrightarrow{BD}$ bisects $\angle ABC$

Before you write out the formal proof, come up with a game plan. Here's one possibility:

- ✔ **Look for congruent triangles.** The congruent triangles should just about jump out at you from this diagram. Think about how you'll show that they're congruent. The triangles share side $\overline{BD}$, giving you one pair of congruent sides. $\overline{BD}$ is an altitude, so that gives you congruent right angles. And because $\overline{BD}$ is a median, $\overline{AD} \cong \overline{CD}$. That does it; you have SAS. (Note that even though you've got right triangles here, you do not use HLR. That wouldn't work because you don't know anything about the hypotenuses.)

- ✔ **Now think about what you have to prove and what you'd need to know to get there.** To conclude that $\overrightarrow{BD}$ bisects $\angle ABC$, you need $\angle ABD \cong \angle CBD$ in the second-to-last line. And how will you get that? Why, with CPCTC, of course!

Here's the two-column proof:

Statements	Reasons
1) $\overline{BD}$ is a median of $\triangle ABC$	1) Given.
2) D is the midpoint of $\overline{AC}$	2) Definition of median.
3) $\overline{AD} \cong \overline{CD}$	3) Definition of midpoint.
4) $\overline{BD}$ is an altitude of $\triangle ABC$	4) Given.
5) $\overline{BD} \perp \overline{AC}$	5) Definition of altitude (if a segment is an altitude [Statement 4], then it is perpendicular to the triangle's base [Statement 5]).
6) $\angle ADB$ is a right angle $\angle CDB$ is a right angle	6) Definition of perpendicular.
7) $\angle ADB \cong \angle CDB$	7) All right angles are congruent.
8) $\overline{BD} \cong \overline{BD}$	8) Reflexive Property.
9) $\triangle ABD \cong \triangle CBD$	9) SAS (3, 7, 8).
10) $\angle ABD \cong \angle CBD$	10) CPCTC.
11) $\overrightarrow{BD}$ bisects $\angle ABC$	11) Definition of bisect.

Every little step in a proof must be spelled out. For instance, in the preceding proof, you can't go from the idea of a median (line 1) to congruent segments (line 3) in one step — even though it's obvious — because the definition of median says nothing about congruent segments. By the same token, you can't go from the idea of an altitude (line 4) to congruent right angles (line 7) in one step or even two steps. You need three steps to connect the links in this chain of logic: Altitude → perpendicular → right angles → congruent angles.

The Isosceles Triangle Theorems

The earlier sections in this chapter involve *pairs* of congruent triangles. Here, you get two theorems that involve a *single* isosceles triangle. Although you often need these theorems for proofs in which you show that two triangles are congruent, the theorems themselves concern only one triangle.

The following two theorems are based on one simple idea about isosceles triangles that happens to work in both directions:

✔ **If sides, then angles:** If two sides of a triangle are congruent, then the angles opposite those sides are congruent (see Figure 5-7).

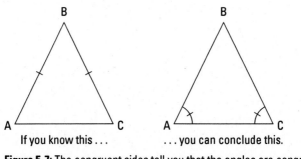

If you know this . . .   . . . you can conclude this.

Figure 5-7: The congruent sides tell you that the angles are congruent.

✔ **If angles, then sides:** If two angles of a triangle are congruent, then the sides opposite those angles are congruent (see Figure 5-8).

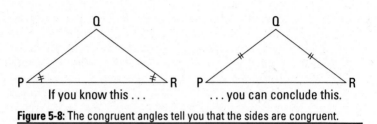

If you know this ...   ... you can conclude this.

Figure 5-8: The congruent angles tell you that the sides are congruent.

Look for isosceles triangles. The two angle-side theorems are critical for solving many proofs, so when you start doing a proof, look at the diagram and identify all triangles that look isosceles. Then make a mental note that you may have to use one of the angle-side theorems for one or more of the isosceles triangles in the diagram. These theorems are incredibly easy to use if you spot all the isosceles triangles (which shouldn't be too hard). But if you fail to notice them, the proof may become impossible.

Here's a proof. Try to work through a game plan and/or a formal proof on your own before reading the ones below.

Given: $\angle P \cong \angle T$

$\overline{PX} \cong \overline{TY}$

$\overline{RX} \cong \overline{RY}$

Prove: $\angle Q \cong \angle S$

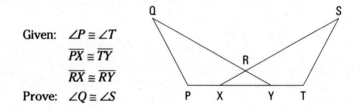

Here's a game plan:

- ✔ **Check the proof diagram for isosceles triangles and pairs of congruent triangles.** This proof's diagram has an isosceles triangle, which is a huge hint that you'll likely use one of the isosceles triangle theorems. You also have a pair of triangles that look congruent (the overlapping ones), which is another huge hint that you'll want to show that they're congruent.

- ✔ **Think about how to finish the proof with a triangle congruence theorem and CPCTC.** You're given the sides of the isosceles triangle, so that gives you congruent angles. You're also given $\angle P \cong \angle T$, so that gives you a second pair of congruent angles. If you can get $\overline{PY} \cong \overline{TX}$, you'd have ASA. And you can get that by adding $\overline{XY}$ to the given congruent segments, $\overline{PX}$ and $\overline{TY}$. You finish with CPCTC.

Statements	Reasons
1) $\overline{RX} \cong \overline{RY}$	1) Given.
2) $\angle RYX \cong \angle RXY$	2) If two sides of a triangle are congruent, then the angles opposite those sides are congruent.
3) $\overline{PX} \cong \overline{TY}$	3) Given.
4) $\overline{PY} \cong \overline{TX}$	4) If a segment is added to two congruent segments, then the sums are congruent.
5) $\angle P \cong \angle T$	5) Given.
6) $\triangle PQY \cong \triangle TSX$	6) ASA (2, 4, 5).
7) $\angle Q \cong \angle S$	7) CPCTC.

The Two Equidistance Theorems

Although congruent triangles are the focus of this chapter, in this section, I give you two theorems that you can often use *instead* of proving triangles congruent. Even though you see congruent triangles in this section's proof diagrams, you don't have to prove the triangles congruent; one of the *equidistance* theorems gives you a shortcut to the *prove* statement.

When doing triangle proofs, be alert for two possibilities: Look for congruent triangles and think about ways to prove them congruent, but at the same time, try to see whether one of the equidistance theorems can get you around the congruent triangle issue.

Determining a perpendicular bisector

The first equidistance theorem tells you that two points determine the perpendicular bisector of a segment. (To "determine" something means to lock in its position, basically to show you where something is.)

Two equidistant points determine the perpendicular bisector:
If two points are each (one at a time) equidistant from the endpoints of a segment, then those points determine the perpendicular bisector of the segment. (Here's an easy way to think
about it: If you have *two* pairs of congruent segments, then
there's a perpendicular bisector.)

The best way to understand this royal mouthful is visually.
Consider the kite-shaped diagram in Figure 5-9.

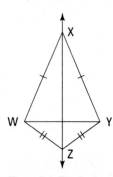

If you know that $\overline{XW} \cong \overline{XY}$ and $\overline{ZW} \cong \overline{ZY}$,
then you can conclude that $\overleftrightarrow{XZ}$ is the
perpendicular bisector of $\overline{WY}$.

Figure 5-9: The first equidistance theorem.

The theorem works like this: If you have one point (like *X*)
that's equally distant from the endpoints of a segment *(W*
and *Y)* and another point (like *Z)* that's also equally distant
from the endpoints, then the two points *(X* and *Z)* determine
the perpendicular bisector of that segment $(\overline{WY})$. You can also
see the meaning of the short form of the theorem in this dia-
gram: If you have *two* pairs of congruent segments $(\overline{XW} \cong \overline{XY}$
and $\overline{ZW} \cong \overline{ZY})$, then there's a perpendicular bisector $(\overleftrightarrow{XZ}$ is the
perpendicular bisector of $\overline{WY})$.

Here's a "*SHORT*" proof that shows how to use the first equidistance theorem as a shortcut that allows you to do the
proof without having to show that the triangles are congruent.

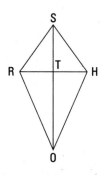

Given: $\overline{SR} \cong \overline{SH}$

$\angle ORT \cong \angle OHT$

Prove: T is the midpoint of $\overline{RH}$

You can do this proof using congruent triangles, but it'd take you about nine steps and you'd have to use two different pairs of congruent triangles.

Statements	Reasons
1) $\angle ORT \cong \angle OHT$	1) Given.
2) $\overline{OR} \cong \overline{OH}$	2) If angles, then sides.
3) $\overline{SR} \cong \overline{SH}$	3) Given.
4) $\overleftrightarrow{SO}$ is the perpendicular bisector of $\overline{RH}$	4) If two points (S and O) are each equidistant from the endpoints of a segment ($\overline{RH}$), then they determine the perpendicular bisector of that segment.
5) $\overline{RT} \cong \overline{TH}$	5) Definition of bisect.
6) T is the midpoint of $\overline{RH}$	6) Definition of midpoint.

Using a perpendicular bisector

With the second equidistance theorem, you use a point on a perpendicular bisector to prove two segments congruent.

A point on the perpendicular bisector is equidistant from the segment's endpoints: If a point is on the perpendicular bisector of a segment, then it's equidistant from the endpoints of the segment. (Here's my abbreviated version: If you have a perpendicular bisector, then there's *one* pair of congruent segments.) Check out Figure 5-10.

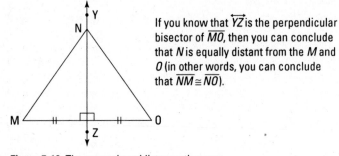

If you know that $\overleftrightarrow{YZ}$ is the perpendicular bisector of $\overline{MO}$, then you can conclude that *N* is equally distant from the *M* and *O* (in other words, you can conclude that $\overline{NM} \cong \overline{NO}$).

Figure 5-10: The second equidistance theorem.

This theorem tells you that if you begin with a segment (like $\overline{MO}$) and its perpendicular bisector (like $\overleftrightarrow{YZ}$) and you have a point on the perpendicular bisector (like *N*), then that point is equally distant from the endpoints of the segment.

Chapter 6

Quadrilaterals

- -

In This Chapter

▶ Crossing the road to get to the other side: Parallel lines and transversals

▶ Tracing the family tree of quadrilaterals

▶ Plumbing the depths of parallelograms, rectangles, and rhombuses

▶ Proving a figure is a parallelogram or other special quadrilateral

- -

*I*n Chapters 4 and 5, you deal with three-sided polygons: triangles. In this chapter, you check out *quadrilaterals*, polygons with four sides. Then, in Chapter 7, you see polygons up to a gazillion sides. Totally exciting, right?

Parallel Line Properties

Most of the quadrilaterals you'll deal with have parallel sides, so let's begin with some info on parallel lines.

Parallel lines with a transversal

Check out Figure 6-1, which shows three lines that kind of resemble a giant not-equal sign. The two horizontal lines are parallel, and the third line that crosses them is called a *transversal*. As you can see, the three lines form eight angles.

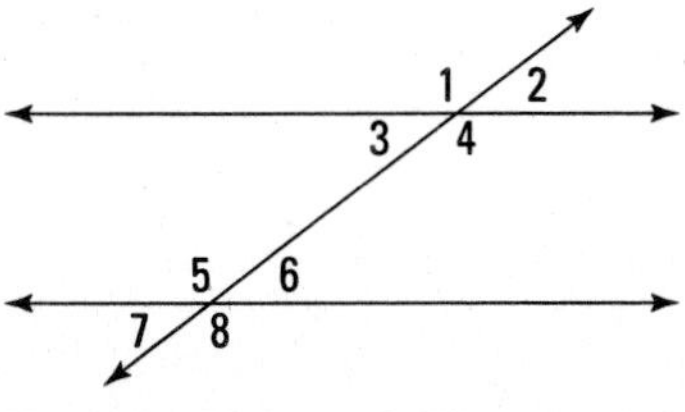

Figure 6-1: Two parallel lines, one transversal, and eight angles.

The eight angles formed by parallel lines and a transversal are either congruent or supplementary. The following theorems tell you how various pairs of angles relate to each other.

Proving that angles are congruent: If a transversal intersects two parallel lines, then the following angles are congruent:

- ✔ **Alternate interior angles:** The pair of angles 3 and 6 (as well as 4 and 5) are *alternate interior angles.* These angle pairs are on opposite (alternate) sides of the transversal and are in between (in the interior of) the parallel lines.

- ✔ **Alternate exterior angles:** Angles 1 and 8 (and angles 2 and 7) are called *alternate exterior angles.*

- ✔ **Corresponding angles:** The pair of angles 1 and 5 (also 2 and 6, 3 and 7, and 4 and 8) are *corresponding angles.* Angles 1 and 5 are corresponding because each is in the same position (the upper left-hand corner) in its group of four angles.

Proving that angles are supplementary: If a transversal intersects two parallel lines, then the following angles are supplementary (see Figure 6-1 again):

- ✔ **Same-side interior angles:** Angles 3 and 5 (and 4 and 6) are on the same side of the transversal and are in the interior of the parallel lines, so they're called (ready for a shock?) *same-side interior angles.*

- ✔ **Same-side exterior angles:** Angles 1 and 7 (and 2 and 8) are called *same-side exterior angles.*

You can sum up the definitions and theorems above with the following simple idea. When you have two parallel lines cut by a transversal, you get four acute angles and four obtuse angles (except when you get eight right angles). All the acute angles are congruent, all the obtuse angles are congruent, and each acute angle is supplementary to each obtuse angle.

Proving that lines are parallel: All the theorems in this section work in reverse. You can use the following theorems to prove that lines are parallel. That is, two lines are parallel if they're cut by a transversal such that

- ✔ Two corresponding angles are congruent.

- ✔ Two alternate interior angles are congruent.

- ✔ Two alternate exterior angles are congruent.

> ✔ Two same-side interior angles are supplementary.
>
> ✔ Two same-side exterior angles are supplementary.

The transversal theorems

Let's take a look at some of the theorems in action: Given that lines m and n are parallel, find the measure of $\angle 1$.

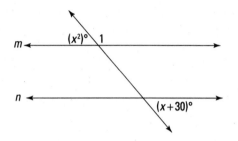

The $(x^2)°$ angle and the $(x + 30)°$ angle are alternate exterior angles and are therefore congruent. Set them equal to each other and solve for x:

$$x^2 = x + 30$$

Set equal to zero: $x^2 - x - 30 = 0$

Factor: $(x - 6)(x + 5) = 0$

Use Zero Product Property: $x - 6 = 0 \quad \text{or} \quad x + 5 = 0$

$$x = 6 \quad \text{or} \quad x = -5$$

This equation has two solutions, so take them one at a time and plug them into the x's in the alternate exterior angles. Plugging $x = 6$ into x^2 gives you 36° for that angle. And because $\angle 1$ is its supplement, $\angle 1$ must be 180° – 36°, or 144°. The $x = -5$ solution gives you 25° for the x^2 angle and 155° for $\angle 1$. So 144° and 155° are your answers for $\angle 1$.

When you get two solutions (such as $x = 6$ and $x = -5$) in a problem like this, you *do not* plug one of them into one of the x's (like $6^2 = 36$) and the other solution into the other x (like –5 + 30 = 25). You have to plug one of the solutions into *all* x's, giving you one result for both angles ($6^2 = 36$ and 6 + 30 = 36); then you have to separately plug the other solution into *all* x's, giving you a second result for both angles ($(-5)^2 = 25$ and –5 + 30 = 25).

Angles and segments can't have measures or lengths that are zero or negative. Make sure that each solution for *x* produces *positive* answers for *all* the angles or segments in a problem. (In the preceding problem, you should check both the $(x^2)°$ angle and the $(x + 30)°$ angle with each solution for *x*.) If a solution makes any angle or segment in the diagram zero or negative, it must be rejected even if the angles or segments you care about end up being positive. However, *do not* reject a solution just because *x* is zero or negative: *x* can be zero or negative as long as the angles and segments are positive ($x = -5$, for example, works just fine in the problem above).

Now here's a proof that uses some of the transversal theorems:

Given: $\overline{LK} \cong \overline{HI}$

$\overline{LK} \parallel \overline{HI}$

$\overline{GK} \cong \overline{JH}$

Prove: $\overline{LJ} \parallel \overline{GI}$

Statements	Reasons
1) $\overline{LK} \cong \overline{HI}$	1) Given.
2) $\overline{LK} \parallel \overline{HI}$	2) Given.
3) $\angle K \cong \angle H$	3) If lines are parallel, then alternate interior angles are congruent.
4) $\overline{GK} \cong \overline{JH}$	4) Given.
5) $\overline{JK} \cong \overline{GH}$	5) If a segment ($\overline{GJ}$) is subtracted from two congruent segments, then the differences are congruent.
6) $\triangle JKL \cong \triangle GHI$	6) SAS (1, 3, 5).
7) $\angle LJK \cong \angle IGH$	7) CPCTC.
8) $\overline{LJ} \parallel \overline{GI}$	8) If alternate exterior angles are congruent, then lines are parallel.

You may want to extend the lines in transversal problems. Doing so can help you see how the angles are related.

For instance, if you have a hard time seeing that ∠K and ∠H are indeed alternate interior angles (for step 3 of the proof), rotate the figure (or tilt your head) until the parallel segments $\overline{LK}$ and $\overline{HI}$ are horizontal; then extend $\overline{LK}$, $\overline{HI}$, and $\overline{HK}$ in both directions, turning them into lines (you know, with the little arrows). After doing that, you're looking at the familiar parallel-line scheme shown in Figure 6-1.

The Seven Special Quadrilaterals

In this section and the next, you find out about the seven quadrilaterals. Some are surely familiar to you, and some may not be so familiar. Check out the following definitions and the quadrilateral family tree in Figure 6-2.

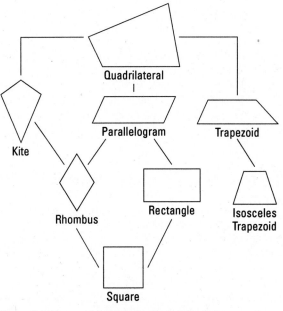

Figure 6-2: The royal family tree of quadrilaterals.

If you know what the quadrilaterals look like, their definitions should make sense and be pretty easy to understand (though the first one is a bit of a mouthful).

- **Kite:** A quadrilateral in which two disjoint pairs of consecutive sides are congruent. ("Disjoint pairs" means that one side can't be used in both pairs.)

- **Parallelogram:** A quadrilateral that has two pairs of parallel sides.

- **Rhombus:** A quadrilateral with four congruent sides. A rhombus is both a kite and a parallelogram.

- **Rectangle:** A quadrilateral with four right angles. A rectangle is a type of parallelogram.

- **Square:** A quadrilateral with four congruent sides and four right angles. A square is both a rhombus and a rectangle.

- **Trapezoid:** A quadrilateral with exactly one pair of parallel sides. (The parallel sides are called *bases*.)

- **Isosceles trapezoid:** A trapezoid in which the nonparallel sides (the *legs*) are congruent.

In the hierarchy of quadrilaterals shown in Figure 6-2, a quadrilateral below another on the family tree is a special case of the one above it. A rectangle, for example, is a special case of a parallelogram. Thus, you can say that a rectangle is a parallelogram but not that a parallelogram is a rectangle (because a parallelogram is only *sometimes* a rectangle).

Working with Auxiliary Lines

The following proof introduces you to a new idea: adding a line or segment (called an *auxiliary line*) to a proof diagram to help you do the proof.

Auxiliary lines often create congruent triangles, or they intersect existing lines at right angles. So if you're stumped by a proof, ask yourself whether drawing an auxiliary line could get you one of those things.

Two points determine a line: When you draw in an auxiliary line, just write something like "Draw $\overline{AB}$" in the statement column;

then use the following postulate in the reason column: Two points determine a line (or ray or segment).

Here's an example proof. (For this proof to make sense, you're not allowed to use the fact that opposite sides of a parallelogram are congruent; I'll get to that in the next section).

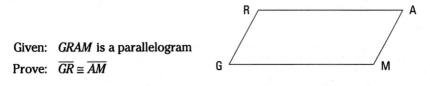

Given: *GRAM* is a parallelogram

Prove: $\overline{GR} \cong \overline{AM}$

You might come up with a game plan like the following:

- ✔ **Take a look at the givens.** The only thing you can conclude from the single given is that the sides of *GRAM* are parallel (using the definition of a parallelogram). But it doesn't seem like you can go anywhere from there.

- ✔ **Jump to the end of the proof.** What could be the justification for the final statement, $\overline{GR} \cong \overline{AM}$? At this point, no justification seems possible, so put on your thinking cap.

- ✔ **Consider drawing an auxiliary line.** If you draw $\overline{RM}$, as shown in Figure 6-3, you get triangles that look congruent. And if you could show that they're congruent, the proof could then end with CPCTC.

- ✔ **Show the triangles congruent.** To show that the triangles are congruent, you use $\overline{RM}$ as a transversal. First use it with parallel sides $\overline{RA}$ and $\overline{GM}$; that gives you congruent, alternate interior angles *GMR* and *ARM*. Then use $\overline{RM}$ with parallel sides $\overline{GR}$ and $\overline{MA}$; that gives you two more congruent, alternate interior angles, *GRM* and *AMR*. These two pairs of congruent angles, along with side $\overline{RM}$ (which is congruent to itself by the Reflexive Property), prove the triangles congruent with ASA. That does it.

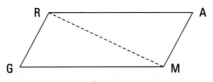

Figure 6-3: Connecting two points creates triangles you can use.

Statements	Reasons
1) *GRAM* is a parallelogram	1) Given.
2) Draw $\overline{RM}$	2) Two points determine a segment.
3) $\overline{RA} \parallel \overline{GM}$	3) Definition of parallelogram.
4) $\angle GMR \cong \angle ARM$	4) If two parallel lines ($\overleftrightarrow{RA}$ and $\overleftrightarrow{GM}$) are cut by a transversal ($\overline{RM}$), then alternate interior angles are congruent.
5) $\overline{GR} \parallel \overline{MA}$	5) Definition of parallelogram.
6) $\angle GRM \cong \angle AMR$	6) Same as Reason 4, but this time $\overleftrightarrow{GR}$ and $\overleftrightarrow{MA}$ are the parallel lines.
7) $\overline{RM} \cong \overline{MR}$	7) Reflexive Property.
8) $\triangle GRM \cong \triangle AMR$	8) ASA (4, 7, 6).
9) $\overline{GR} \cong \overline{AM}$	9) CPCTC.

A good way to spot congruent alternate interior angles in a diagram is to look for pairs of so-called *Z-angles.* Look for a Z or backward Z — or a stretched-out Z or backward Z — as shown in Figures 6-4 and 6-5.

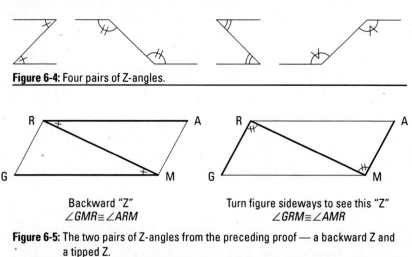

Figure 6-4: Four pairs of Z-angles.

Backward "Z"
$\angle GMR \cong \angle ARM$

Turn figure sideways to see this "Z"
$\angle GRM \cong \angle AMR$

Figure 6-5: The two pairs of Z-angles from the preceding proof — a backward Z and a tipped Z.

The Properties of Quadrilaterals

The *properties* of the quadrilaterals are simply the things that are true about them. The properties of a particular quadrilateral concern its

- ✓ **Sides:** Are they congruent? Parallel?
- ✓ **Angles:** Are they congruent? Supplementary? Right?
- ✓ **Diagonals:** Are they congruent? Perpendicular? Do they bisect each other? Do they bisect the angles whose vertices they meet?

I present a total of about 30 quadrilateral properties, which may seem like a lot to memorize. No worries. You don't have to rely solely on memorization. Here's a great tip that makes learning the properties a snap.

If you can't remember whether something is a property of some quadrilateral, just sketch the quadrilateral in question. If the thing looks true, it's probably a property; if it doesn't look true, it's not a property. (This method is almost foolproof, but it's a bit un-math-teacherly of me to say it — so don't quote me, or I might get in trouble with the math police.)

Properties of the parallelogram

The parallelogram has the following properties:

- ✓ Opposite sides are parallel by definition.
- ✓ Opposite sides are congruent.
- ✓ Opposite angles are congruent.
- ✓ Consecutive angles are supplementary.
- ✓ The diagonals bisect each other.

If you just look at a parallelogram, the things that look true (namely, the things on this list) *are* true and are thus properties, and the things that don't look like they're true aren't properties.

If you draw a picture to help you figure out a quadrilateral's properties, make your sketch as general as possible. For instance, as you sketch your parallelogram, make sure it's not almost a rhombus (with four sides that are almost congruent) or almost a rectangle (with four angles close to right angles).

Imagine that you can't remember the properties of a parallelogram. You could just sketch one (as in Figure 6-6) and run through all things that might be properties.

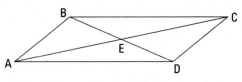

Figure 6-6: A run-of-the-mill parallelogram.

The following tables concern questions about what might or might not be properties of a parallelogram (refer to Figure 6-6).

Table 6-1   Questions about Sides of Parallelograms

Do Any Sides Appear to Be . . .	Answer
Congruent?	Yes, opposite sides look congruent, and that's a property. But adjacent sides don't look congruent, and that's not a property.
Parallel?	Yes, opposite sides look parallel (and of course, you know this property if you know the definition of a parallelogram).

Table 6-2   Questions about Angles of Parallelograms

Do Any Angles Appear to Be . . .	Answer
Congruent?	Yes, opposite angles look congruent, and that's a property. (Angles A and C appear to be about 45°, and angles B and D look like about 135°).
Supplementary?	Yes, consecutive angles (like angles A and B) look like they're supplementary, and that's a property.
Right angles?	Obviously not, and that's not a property.

Table 6-3   Questions about Parallelogram Diagonals

Do the Diagonals Appear to Be . . .	Answer
Congruent?	Not even close (in Figure 6-6, one is roughly twice as long as the other, which surprises most people; measure them if you don't believe me!) — not a property.
Perpendicular?	Not even close; not a property.
Bisecting each other?	Yes, each one seems to cut the other in half, and that's a property.
Bisecting the angles whose vertices they meet?	No. At a quick glance, you might think that $\angle A$ (or $\angle C$) is bisected by diagonal $\overline{AC}$, but if you look closely, you see that $\angle BAC$ is actually about twice as big as $\angle DAC$. And of course, diagonal $\overline{BD}$ doesn't come close to bisecting $\angle B$ or $\angle D$. Not a property.

Look at your sketch carefully. When I show students a parallelogram like the one in Figure 6-6 and ask them whether the diagonals look congruent, they often tell me that they do despite the fact that one is literally twice as long as the other! So when asking yourself whether a potential property looks true, don't just take a quick glance at the quadrilateral, and don't let your eyes play tricks on you. Look at the segments or angles in question very carefully.

The sketching-quadrilaterals method and the questions in the preceding three tables bring me to an important tip about mathematics in general.

Whenever possible, bolster your memorization of rules, formulas, concepts, and so on by trying to see *why* they're true or *why* they make sense. Not only does this make the ideas easier to learn, but it also helps you see connections to other ideas, and that fosters a deeper understanding of mathematics.

Properties of the three special parallelograms

Figure 6-7 shows you the three *special* parallelograms, so-called because they're, as mathematicians say, *special cases* of the parallelogram. (In addition, the square is a special case or type of both the rectangle and the rhombus.) The three-level hierarchy you see with *parallelogram → rectangle → square* or *parallelogram → rhombus → square* in the quadrilateral family tree (Figure 6-2) works just like *mammal → dog → Dalmatian*. A dog is a special type of a mammal, and a Dalmatian is a special type of a dog.

Before reading the properties that follow, try figuring them out on your own. Using the shapes in Figure 6-7, run down the list of possible properties from the beginning of "The Properties of Quadrilaterals," asking yourself whether they look like they're true for the rhombus, the rectangle, and the square.

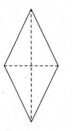
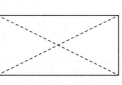

Rhombus   Rectangle   Square

Figure 6-7: The two kids and one grandkid (the square) of the parallelogram.

Here are the properties of the rhombus, rectangle, and square. Note that because these three quadrilaterals are all parallelograms, their properties include the parallelogram properties.

✔ **The rhombus has the following properties:**

- All the properties of a parallelogram apply.

- All sides are congruent by definition.

- The diagonals bisect the angles.

- The diagonals are perpendicular bisectors of each other.

✔ **The rectangle has the following properties:**

- All the properties of a parallelogram apply.

- All angles are right angles by definition.

- The diagonals are congruent.

✔ **The square has the following properties:**

- All the properties of a rhombus apply.
- All the properties of a rectangle apply.
- All sides are congruent by definition.
- All angles are right angles by definition.

Let's try a problem: Find the perimeter of rhombus *RHOM*.

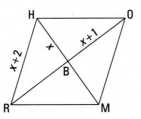

Here's the solution: All the sides of a rhombus are congruent, so *HO* equals $x + 2$. And because the diagonals of a rhombus are perpendicular, $\triangle HBO$ is a right triangle. You finish with the Pythagorean Theorem:

$$a^2 + b^2 = c^2$$
$$(HB)^2 + (BO)^2 = (HO)^2$$
$$x^2 + (x+1)^2 = (x+2)^2$$
$$x^2 + x^2 + 2x + 1 = x^2 + 4x + 4$$

Combine like terms and set equal to zero: $\quad x^2 - 2x - 3 = 0$

Factor: $\qquad\qquad\qquad\qquad\qquad (x-3)(x-1) = 0$

Use Zero Product Property:

$$x - 3 = 0 \quad \text{or} \quad x + 1 = 0$$
$$x = 3 \qquad \text{or} \quad x = -1$$

You can reject $x = -1$ because that would result in $\triangle HBO$ having legs with lengths of -1 and 0. So x equals 3, which gives $\overline{HO}$ a length of 5. Because rhombuses have four congruent sides, *RHOM* has a perimeter of $4 \cdot 5$, or 20 units.

Properties of the kite

Check out the kite in Figure 6-8 and try to figure out its properties before reading the list that follows.

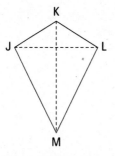

Figure 6-8: A mathematical kite that looks ready for flying.

The properties of the kite are as follows:

✔ Two disjoint pairs of consecutive sides are congruent by definition ($\overline{JK} \cong \overline{LK}$ and $\overline{JM} \cong \overline{LM}$). *Note: Disjoint* means that one side can't be used in both pairs — the two pairs are totally separate.

✔ The diagonals are perpendicular.

✔ One diagonal ($\overline{KM}$, the *main diagonal*) is the perpendicular bisector of the other diagonal ($\overline{JL}$, the *cross diagonal*). (The terms "main diagonal" and "cross diagonal" are quite useful, but don't look for them in other geometry books because I made them up.)

✔ The main diagonal bisects a pair of opposite angles ($\angle K$ and $\angle M$).

✔ The opposite angles at the endpoints of the cross diagonal are congruent ($\angle J$ and $\angle L$).

The last three properties are called the *half properties* of the kite.

Grab an energy drink and get ready for another proof. Due to space considerations, I'm going to skip the game plan this time. You're on your own — egad!

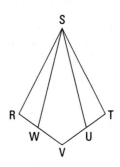

Given: $RSTV$ is a kite with $\overline{RS} \cong \overline{TS}$

$\overline{WV} \cong \overline{UV}$

Prove: $\overline{WS} \cong \overline{US}$

Statements	Reasons
1) $RSTV$ is a kite with $\overline{RS} \cong \overline{TS}$	1) Given.
2) $\overline{RV} \cong \overline{TV}$	2) A kite has two disjoint pairs of congruent sides.
3) $\overline{WV} \cong \overline{UV}$	3) Given.
4) $\overline{RW} \cong \overline{TU}$	4) If two cong. segments ($\overline{WV}$ and $\overline{UV}$) are subtracted from two other cong. segments ($\overline{RV}$ and $\overline{TV}$), then the differences are congruent.
5) $\angle R \cong \angle T$	5) The opposite angles at the endpoints of the cross diagonal are congruent.
6) $\triangle SRW \cong \triangle STU$	6) SAS (1, 5, 4).
7) $\overline{WS} \cong \overline{US}$	7) CPCTC.

Properties of the trapezoid and the isosceles trapezoid

Practice your picking-out-properties proficiency with the trapezoid and isosceles trapezoid in Figure 6-9. *Remember:* What looks true is likely true, and what doesn't, isn't.

Figure 6-9: A trapezoid (on the left) and an isosceles trapezoid (on the right).

✔ **The properties of the trapezoid are as follows:**

- The bases are parallel by definition.
- Each lower base angle is supplementary to the upper base angle on the same side.

✔ **The properties of the isosceles trapezoids are as follows:**

- The properties of trapezoids apply by definition
- The legs are congruent by definition.
- The lower base angles are congruent.
- The upper base angles are congruent.
- Any lower base angle is supplementary to any upper base angle.
- The diagonals are congruent.

Proving That You've Got a Particular Quadrilateral

The last two sections told you all about seven different quadrilaterals — their definitions, their properties, what they look like, and where they fit on the family tree. Here, I fill you in on proving that a given quadrilateral qualifies as one of those particular types.

Proving you've got a parallelogram

The five methods for proving that a quadrilateral is a parallelogram are among the most important proof methods in this section. One reason they're important is that you often have to prove that a quadrilateral is a parallelogram before going on to prove that it's one of the special parallelograms (a rectangle, a rhombus, or a square).

Five ways to prove that a quadrilateral is a parallelogram: There are five ways to prove that a quadrilateral is a parallelogram. The first four are the converses of parallelogram properties. Make sure you remember the oddball fifth one — which isn't the converse of a property — because it often comes in handy:

✔ If both pairs of opposite sides of a quadrilateral are parallel, then it's a parallelogram (reverse of the parallelogram definition).

✔ If both pairs of opposite sides of a quadrilateral are congruent, then it's a parallelogram.

✔ If both pairs of opposite angles of a quadrilateral are congruent, then it's a parallelogram.

✔ If the diagonals of a quadrilateral bisect each other, then it's a parallelogram.

✔ If one pair of opposite sides of a quadrilateral are both parallel and congruent, then it's a parallelogram.

Here's a proof to give you some practice with one of the parallelogram proof methods.

Given: *HEJG* is a parallelogram

∠*DGH* ≅ ∠*FEJ*

Prove: *DEFG* is a parallelogram

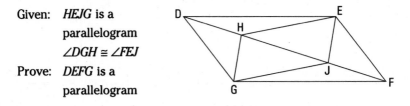

Because all quadrilaterals (except for the kite) contain parallel lines, be on the lookout for opportunities to use the parallel-line theorems from early in this chapter. And always keep your eyes peeled for congruent triangles.

Your game plan might go something like this:

✔ **Look for congruent triangles.** This diagram takes the cake for containing congruent triangles — it has six pairs of them! Don't spend much time thinking about them — except the ones that might help you — but at least make a quick mental note that they're there.

✔ **Consider the givens.** The given congruent angles, which are parts of △*DGH* and △*FEJ* are a huge hint that you should try to show these triangles congruent. You have those congruent angles and the congruent sides $\overline{HG}$ and $\overline{EJ}$ from parallelogram *HEJG* so you need only one more pair of congruent sides or angles to use SAS or ASA.

✔ **Think about the end of the proof.** To prove that *DEFG* is a parallelogram, it would help to know that $\overline{DG} \cong \overline{EF}$ so you'd like to be able to prove the triangles congruent and then get $\overline{DG} \cong \overline{EF}$ by CPCTC. That eliminates the SAS option for proving the triangles congruent because to use SAS, you'd need to know that $\overline{DG} \cong \overline{EF}$ — the very thing you're trying to get with CPCTC. (And if you knew $\overline{DG} \cong \overline{EF}$ there'd be no point to showing that the triangles are congruent, anyway.) So you should try the other option: proving the triangles congruent with ASA.

The second angle pair you'd need for ASA consists of $\angle DHG$ and $\angle FJE$. They're congruent because they're alternate exterior angles using parallel lines $\overleftrightarrow{HG}$ and $\overleftrightarrow{EJ}$ and transversal $\overleftrightarrow{DF}$. Okay, so the triangles are congruent by ASA, and then you get $\overline{DG} \cong \overline{EF}$ by CPCTC. You're on your way.

✔ **Consider parallelogram proof methods.** You now have one pair of congruent sides of *DEFG*. Two of the parallelogram proof methods use a pair of congruent sides. To complete one or the other of these methods, you need to show one of the following:

- That the other pair of opposite sides are congruent

- That $\overline{DG}$ and $\overline{EF}$ are parallel as well as congruent

Ask yourself which approach looks easier or quicker. Showing $\overline{DE} \cong \overline{GF}$ would probably require showing a second pair of triangles congruent, and that looks like it'd take a few more steps, so try the other tack.

Can you show $\overline{DG} \parallel \overline{EF}$? Sure, with one of the parallel-line theorems. Because angles *GDH* and *EFJ* are congruent (by CPCTC), you can finish by using those angles as congruent alternate interior angles, or Z-angles, to give you $\overline{DG} \parallel \overline{EF}$. That's a wrap!

Now take a look at the formal proof:

Statements	Reasons
1) *HEJG* is a parallelogram	1) Given.
2) $\overline{HG} \cong \overline{EJ}$	2) Opposite sides of a parallelogram are congruent.
3) $\overline{HG} \parallel \overline{EJ}$	3) Opposite sides of a parallelogram are parallel.

4) ∠*DHG* ≅ ∠*FJE*	4) If lines are parallel, then alternate exterior angles are congruent.
5) ∠*DGH* ≅ ∠*FEJ*	5) Given.
6) △*DGH* ≅ △*FEJ*	6) ASA (4, 2, 5).
7) $\overline{DG}$ ≅ $\overline{EF}$	7) CPCTC.
8) ∠*GDH* ≅ ∠*EFJ*	8) CPCTC.
9) $\overline{DG}$ ∥ $\overline{EF}$	9) If alternate interior angles are congruent (∠*GDH* and ∠*EFJ*), then lines are parallel.
10) *DEFG* is a parallelogram	10) If one pair of opposite sides of a quadrilateral are both parallel and congruent, then the quadrilateral is a parallelogram (lines 9 and 7).

Proving that you've got a rectangle, rhombus, or square

Three ways to prove that a quadrilateral is a rectangle: Note that the second and third methods require that you first show (or be given) that the quadrilateral in question is a parallelogram:

✔ If all angles in a quadrilateral are right angles, then it's a rectangle (reverse of the rectangle definition).

✔ If the diagonals of a parallelogram are congruent, then it's a rectangle.

✔ If a parallelogram contains a right angle, then it's a rectangle.

Six ways to prove that a quadrilateral is a rhombus: You can use the following six methods to prove that a quadrilateral is a rhombus. The last three methods require that you first show (or be given) that you've got a parallelogram:

✔ If all sides of a quadrilateral are congruent, then it's a rhombus (reverse of the rhombus definition).

✔ If the diagonals of a quadrilateral bisect all the angles, then it's a rhombus.

✔ If the diagonals of a quadrilateral are perpendicular bisectors of each other, then it's a rhombus.

✔ If two consecutive sides of a parallelogram are congruent, then it's a rhombus.

✔ If either diagonal of a parallelogram bisects two angles, then it's a rhombus.

✔ If the diagonals of a parallelogram are perpendicular, then it's a rhombus.

Four methods to prove that a quadrilateral is a square: In the last three of these methods, you first have to prove (or be given) that the quadrilateral is a rectangle, rhombus, or both:

✔ If a quadrilateral has four congruent sides and four right angles, then it's a square (reverse of the square definition).

✔ If two consecutive sides of a rectangle are congruent, then it's a square.

✔ If a rhombus contains a right angle, then it's a square.

✔ If a quadrilateral is both a rectangle and a rhombus, then it's a square.

Proving that you've got a kite

Two ways to prove that a quadrilateral is a kite: Proving that a quadrilateral is a kite is pretty easy. Usually, all you have to do is use congruent triangles or isosceles triangles:

✔ If two disjoint pairs of consecutive sides of a quadrilateral are congruent, then it's a kite (reverse of the kite definition).

✔ If one of the diagonals of a quadrilateral is the perpendicular bisector of the other, then it's a kite.

When you're trying to prove that a quadrilateral is a kite, the following tips may come in handy:

✔ **Check the diagram for congruent triangles.** Don't fail to spot triangles that look congruent and to consider how using CPCTC might help you.

✔ **Keep the first equidistance theorem in mind.** When you have to prove that a quadrilateral is a kite, you might have to use the equidistance theorem in which two points determine a perpendicular bisector.

✔ **Draw in diagonals.** One of the methods for proving that a quadrilateral is a kite involves diagonals, so if the diagram lacks either of the two diagonals, try drawing in one or both of them.

Now get ready for a proof:

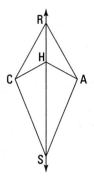

Given: $\overleftrightarrow{RS}$ bisects ∠*CRA* and ∠*CHA*

Prove: *CRAS* is a kite

Here's how your game plan might work for this proof.

✔ **Note that one of the kite's diagonals is missing.** Draw in the missing diagonal, $\overline{CA}$.

✔ **Check the diagram for congruent triangles.** After drawing in $\overline{CA}$, there are six pairs of congruent triangles. The two triangles most likely to help you are △*CRH* and △*ARH*.

✔ **Prove the triangles congruent.** You can use ASA.

✔ **Use the equidistance theorem.** Use CPCTC with △*CRH* and △*ARH* to get $\overline{CR} \cong \overline{AR}$ and $\overline{CH} \cong \overline{AH}$. Then, using the equidistance theorem, those two pairs of congruent sides (with points R and H) determine the perpendicular bisector of the diagonal you drew in. Over and out.

I'm skipping the formal proof, but give it a try. It should take you about 12 steps.

Chapter 7

Polygon Formulas

*I*n this chapter, you take a break from proofs and move on to problems that have a *little* more to do with the real world. I emphasize *little* because the shapes you deal with here — such as trapezoids, hexagons, octagons, and yep, even pentadecagons (15 sides) — aren't exactly things you encounter outside of math class on a regular basis. But at least the concepts you work with here — the length and size and shape of polygons — are fairly ordinary things. For nearly everyone, relating to visual, real-world things like this is easier than relating to proofs, which are more in the realm of pure mathematics.

The Area of Quadrilaterals

I'm sure you've had to calculate the area of a square or rectangle before, whether it was in a math class or in some more practical situation, such as when you wanted to know the area of a room in your house. In this section, you see the square and rectangle formulas again, and you also get some new, gnarlier formulas you may not have seen before.

Quadrilateral area formulas

Here are the five area formulas for the seven special quadrilaterals. There are only five formulas because some of them do double duty — for example, you can calculate the area of a rhombus with the kite formula.

Quadrilateral area formulas:

- ✔ Area $_{\text{Rectangle}}$ = base · height (or length · width)

- ✔ Area $_{\text{Parallelogram}}$ = base · height (because a rhombus is a type of parallelogram, you can use this formula for a rhombus)

- ✔ Area $_{\text{Kite}}$ = $\frac{1}{2}$ diagonal$_1$ · diagonal$_2$, or $\frac{1}{2}d_1d_2$ (a rhombus is also a type of kite, so you can use this formula for a rhombus as well)

- ✔ Area $_{\text{Square}}$ = side^2, or $\frac{1}{2}$ diagonal^2 (this second formula works because a square is a type of kite)

- ✔ Area $_{\text{Trapezoid}}$ = $\dfrac{\text{base}_1 + \text{base}_2}{2}$ · height

  = median · height

Note: The *median* of a trapezoid is the segment that connects the midpoints of the legs. Its length equals the average of the lengths of the bases.

Why the formulas work

The area formulas for the parallelogram, kite, and trapezoid are based on the area of a rectangle. The following figures show you how each of these three quadrilaterals relates to a rectangle, and the following list gives you the details:

- ✔ **Parallelogram:** In Figure 7-1, if you cut off the little triangle on the left and fill it in on the right, the parallelogram becomes a rectangle (and the area obviously hasn't changed). This rectangle has the same base and height as the original parallelogram. The area of the rectangle is *base · height,* so that formula gives you the area of the parallelogram as well. If you don't believe me (even though you should by now) you can try this yourself by cutting out a paper parallelogram and snipping off the triangle as shown in Figure 7-1.

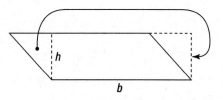

Figure 7-1: The relationship between a parallelogram and a rectangle.

✔ **Kite:** Figure 7-2 shows that the kite has half the area of the rectangle drawn around it (this follows from the fact that $\triangle 1 \cong \triangle 2$, $\triangle 3 \cong \triangle 4$, and so on). You can see that the length and width of the large rectangle are the same as the lengths of the diagonals of the kite. The area of the rectangle (length · width) thus equals $d_1 d_2$, and because the kite has half that area, its area is $\frac{1}{2} d_1 d_2$.

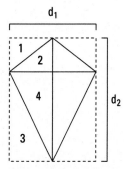

Figure 7-2: The kite takes up half of each of the four small rectangles and thus is half the area of the large rectangle.

✔ **Trapezoid:** If you cut off the two triangles and move them as I show you in Figure 7-3, the trapezoid becomes a rectangle. This rectangle has the same height as the trapezoid, and its base equals the median *(m)* of the trapezoid. Thus, the area of the rectangle (and therefore the trapezoid as well) is *median · height.*

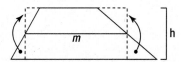

Figure 7-3: The relationship between a trapezoid and a rectangle.

Trying a few area problems

The key for many quadrilateral area problems is to draw altitudes and other perpendicular segments on the diagram. Doing
so creates one or more right triangles, which allows you to use
the Pythagorean Theorem or your knowledge of special right
triangles, such as the 45°-45°-90° and 30°-60°-90° triangles.

Locating special right triangles in parallelograms

Find the area of parallelogram *ABCD* in Figure 7-4.

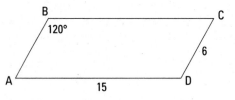

Figure 7-4: Use a 30°-60°-90° triangle to find the area of this parallelogram.

When you see a 120° angle in a problem, a 30°-60°-90° triangle
is likely lurking somewhere in the problem. (Of course, a 30°
or 60° angle is a dead giveaway of a 30°-60°-90° triangle.) And if
you see a 135° angle, a 45°-45°-90° triangle is likely lurking.

To get started, draw in the height of the parallelogram
straight down from *B* to base $\overline{AD}$ to form a right triangle
as shown in Figure 7-5.

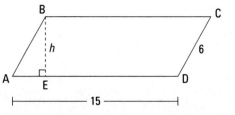

Figure 7-5: Drawing in the height creates a right triangle.

Consecutive angles in a parallelogram are supplementary.
Angle *ABC* is 120°, so ∠*A* is 60° and △*ABE* is thus a 30°-60°-
90° triangle. Now, if you know the ratio of the lengths of the
sides in a 30°-60°-90° triangle, $x : x\sqrt{3} : 2x$, the rest is a snap.

$\overline{AB}$ (the $2x$ side) equals $\overline{CD}$ and is thus 6. Then $\overline{AE}$ (the x side) is half of that, or 3; $\overline{BE}$ (the $x\sqrt{3}$ side) is therefore $3\sqrt{3}$. Here's the finish with the area formula:

$$\text{Area}_{\text{Parallelogram}} = b \cdot h$$
$$= 15 \cdot 3\sqrt{3}$$
$$= 45\sqrt{3} \approx 77.9 \text{ units}^2$$

Using triangles and ratios in a rhombus problem

Now for a rhombus problem: Find the area of rhombus *RHOM* given that $\overline{MB}$ is 6 and that the ratio of $\overline{RB}$ to $\overline{BH}$ is 4 : 1 (see Figure 7-6).

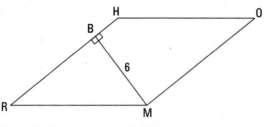

Figure 7-6: Find the area of this rhombus.

This one's a bit tricky. You might feel that you're not given enough information to solve it or that you just don't know how to begin. If you ever feel this way when you're in the middle of a problem, I have a great tip for you.

If you get stuck when doing a geometry problem — or any kind of math problem, for that matter — *do something, anything!* Begin anywhere you can: Use the given information or any ideas you have (try simple ideas before more-advanced ones) and write something down. Maybe draw a diagram if you don't have one. *Put something down on paper.* One idea may trigger another, and before you know it, you've solved the problem. This tip is surprisingly effective.

Because the ratio of $\overline{RB}$ to $\overline{BH}$ is 4 : 1, you can give $\overline{RB}$ a length of $4x$ and $\overline{BH}$ a length of x. $\overline{RH}$ is thus $4x + x$, or $5x$, and so is $\overline{RM}$, because all sides of a rhombus are congruent. Now you have a right triangle ($\triangle RBM$) with legs of $4x$ and 6 and a hypotenuse of $5x$, so you can use the Pythagorean Theorem:

$$a^2 + b^2 = c^2$$
$$(4x)^2 + 6^2 = (5x)^2$$
$$16x^2 + 36 = 25x^2$$
$$36 = 9x^2$$
$$4 = x^2$$
$$x = 2 \text{ or } -2$$

Because side lengths must be positive, you reject the answer $x = -2$. The length of the base, $\overline{RH}$, is thus 5(2), or 10. (Triangle *RBM* is your old, familiar friend, a 3-4-5 triangle blown up by a factor of 2.) Now use the parallelogram-rhombus area formula:

$$\text{Area}_{RHOM} = bh$$
$$= 10 \cdot 6$$
$$= 60 \text{ units}^2$$

Drawing in diagonals to find a kite's area

What's the area of kite *KITE* in Figure 7-7?

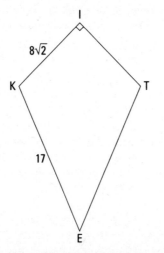

Figure 7-7: A kite with a funky side length.

Draw in diagonals if necessary. For kite and rhombus area problems (and sometimes other quadrilateral problems), the diagonals are almost always necessary for the solution (because they form right triangles). You may have to add them to the figure.

So draw in $\overline{KT}$ and $\overline{IE}$. Use "X" for the point where the diagonals cross. Use some scratch paper, or draw on Figure 7-7; no one will know.

Triangle *KIT* is a right triangle with congruent legs, so it's a 45°- 45°- 90° triangle with sides in the ratio of $x : x : x\sqrt{2}$. The length of the hypotenuse, $\overline{KT}$, thus equals one of the legs times $\sqrt{2}$; that's $8\sqrt{2} \cdot \sqrt{2}$, or 16. $\overline{KX}$ is half of that, or 8.

Triangle *KIX* is another 45°- 45°- 90° triangle ($\overline{IE}$, the kite's main diagonal, bisects opposite angles *KIT* and *KET*, and half of ∠*KIT* is 45°; therefore, $\overline{IX}$, like $\overline{KX}$, is 8. You've got another right triangle, △*KXE*, with a side of 8 and a hypotenuse of 17. I hope that rings a bell! You're looking at an 8-15-17 triangle, so without any work, you see that $\overline{XE}$ is 15. (No bells? No worries. You can get $\overline{XE}$ with the Pythagorean Theorem instead.) Add $\overline{XE}$ to $\overline{IX}$ and you get 8 + 15 = 23 for diagonal $\overline{IE}$.

Now that you know the diagonal lengths, you have what you need to finish. The length of diagonal $\overline{KT}$ is 16, and diagonal $\overline{IE}$ is 23. Plug these numbers into the kite area formula for your final answer:

$$\text{Area}_{KITE} = \frac{1}{2}d_1 d_2$$
$$= \frac{1}{2} \cdot 16 \cdot 23$$
$$= 184 \text{ units}^2$$

The Area of Regular Polygons

In case you've been dying to know how to figure the area of your ordinary, octagonal stop sign, you've come to the right place. In this section, you discover how to find the area of equilateral triangles and other regular polygons.

The polygon area formulas

A *regular polygon* is equilateral (it has equal sides) and equiangular (it has equal angles). To find the area of a regular polygon, you use an *apothem* — a segment that joins the polygon's center to the midpoint of any side and that is perpendicular to that side. ($\overline{HM}$ in upcoming Figure 7-8 is an apothem.)

Area of a regular polygon: Use the following formula to find the area of a regular polygon.

$$\text{Area}_{\text{Regular Polygon}} = \frac{1}{2} \text{ perimeter} \cdot \text{apothem, or } \frac{1}{2}pa$$

Note: This formula is usually written as $\frac{1}{2}ap$, but if I do say so myself, the way I've written it, $\frac{1}{2}pa$, is better. I like this way of writing it because the formula is based on the triangle area formula, $\frac{1}{2}bh$: The polygon's perimeter *(p)* is related to the triangle's base *(b)*, and the apothem *(a)* is related to the height *(h)*.

An equilateral triangle is the regular polygon with the fewest possible number of sides. To figure its area, you can use the regular polygon formula; however, it also has its own area formula.

Area of an equilateral triangle: Here's the area formula for an equilateral triangle.

$$\text{Area}_{\text{Equilateral}\triangle} = \frac{s^2\sqrt{3}}{4} \text{ (where } s \text{ is the length of each of the triangle's sides)}$$

Tackling an area problem

Don't tell me about your problems; I've got problems of my own — and here's one of them.

What's the area of a regular hexagon with an apothem of $10\sqrt{3}$?

For hexagons, use 30°- 60°- 90° and equilateral triangles. A regular hexagon can be cut into six equilateral triangles, and an equilateral triangle can be divided into two 30°- 60°- 90° triangles. So if you're doing a hexagon problem, you may want to cut up the figure and use equilateral triangles or 30°- 60°- 90° triangles to help you find the apothem, perimeter, or area.

First, sketch the hexagon with its three diagonals, creating six equilateral triangles. Then draw in an apothem, which goes from the center to the midpoint of a side. Figure 7-8 shows hexagon *EXAGON*.

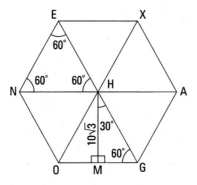

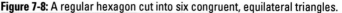
Figure 7-8: A regular hexagon cut into six congruent, equilateral triangles.

Note that the apothem divides $\triangle OHG$ into two 30°- 60°- 90° triangles (halves of an equilateral triangle). The apothem is the long leg (the $x\sqrt{3}$ side) of a 30°- 60°- 90° triangle, so

$$x\sqrt{3} = 10\sqrt{3}$$
$$x = 10$$

$\overline{OM}$ is the short leg (the x side), so its length is 10. $\overline{OG}$ is twice as long, so it's 20. And the perimeter is six times that, or 120.

Now you can finish with either the regular polygon formula or the equilateral triangle formula (multiplied by 6). They're equally easy. Take your pick. Here's what it looks like with the regular polygon formula:

$$\text{Area}_{EXAGON} = \frac{1}{2}pa$$
$$= \frac{1}{2} \cdot 120 \cdot 10\sqrt{3}$$
$$= 600\sqrt{3} \text{ units}^2$$

Angle and Diagonal Formulas

In this section, you get polygon formulas involving — hold onto your hat — angles and diagonals!

Interior and exterior angles

You use two kinds of angles when working with polygons (see Figure 7-9):

- ✔ **Interior angle:** An interior angle of a polygon is an angle inside the polygon at one of its vertices. Angle Q is an interior angle of quadrilateral *QUAD*.

- ✔ **Exterior angle:** An exterior angle of a polygon is an angle outside the polygon formed by one of its sides and the extension of an adjacent side. Angle *ADZ*, $\angle XUG$, and $\angle YUA$ are exterior angles of *QUAD*; vertical angle $\angle XUY$ is *not* an exterior angle of *QUAD*.

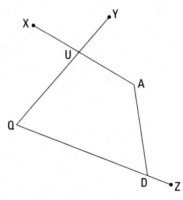

Figure 7-9: Interior and exterior angles.

Interior and exterior angle formulas:

- ✔ The *sum* of the measures of the *interior angles* of a polygon with n sides is $(n-2)180$.

- ✔ The measure of *each interior angle* of an equiangular n-gon is $\dfrac{(n-2)180}{n}$ or $180 - \dfrac{360}{n}$ (the supplement of an exterior angle).

✔ If you count one exterior angle at each vertex, the *sum* of the measures of the *exterior angles* of a polygon is 360°.

✔ The measure of *each exterior angle* of an equiangular *n*-gon is $\frac{360}{n}$.

A polygon angle problem

You can practice the interior and exterior angle formulas in the following three-part problem: Given a regular dodecagon (12 sides),

1. **Find the sum of the measures of its interior angles.**

   Just plug the number of sides (12) into the formula for the sum of the interior angles of a polygon:

   $$\text{Sum of interior angles} = (n-2)180$$
   $$= (12-2)180$$
   $$= 1,800°$$

2. **Find the measure of a single interior angle.**

   This polygon has 12 sides, so it has 12 angles; and because you're dealing with a *regular* polygon, all its angles are congruent. So to find the measure of a single angle, just divide your answer from the first part of the problem by 12. (Note that this is basically the same as using the first formula for a single interior angle.)

   $$\text{Measure of a single interior angle} = \frac{1,800}{12}$$
   $$= 150°$$

3. **Find the measure of a single exterior angle with the exterior angle formula; then check that its supplement, an interior angle, equals the answer you got from part 2 of the problem.**

   First, plug 12 into the oh-so-simple exterior angle formula:

   $$\text{Measure of a single exterior angle} = \frac{360}{12}$$
   $$= 30°$$

Now take the supplement of your answer to find the measure of a single interior angle, and check that it's the same as your answer from part 2:

$$\text{Measure of a single interior angle} = 180 - 30$$
$$= 150°$$

It checks. (And note that this final computation is basically the same thing as using the second formula for a single interior angle.)

Criss-crossing with diagonals

Number of diagonals in an *n*-gon: The number of diagonals that you can draw in an *n*-gon is $\frac{n(n-3)}{2}$.

Here's one last problem for you: If a polygon has 90 diagonals, how many sides does it have?

You know what the formula for the number of diagonals in a polygon is, and you know that the polygon has 90 diagonals, so plug 90 in for the answer and solve for *n*:

$$\frac{n(n-3)}{2} = 90$$
$$n^2 - 3n = 180$$
$$n^2 - 3n - 180 = 0$$
$$(n-15)(n+12) = 0$$

Thus, *n* equals 15 or –12. But because a polygon can't have a negative number of sides, *n* must be 15. So you have a 15-sided polygon (a pentadecagon, in case you're curious).

Chapter 8

Similarity

* *

In This Chapter

▶ Sizing up similar figures

▶ Doing similar triangle proofs

▶ Scoping out theorems about proportionality

* *

*Y*ou know the meaning of the word *similar* in everyday speech. In geometry, it has a related but more technical meaning. Two figures are *similar* if they have exactly the same shape.

Similar Figures

In this section, I cover the formal definition of similarity, how similar figures are named, and how they're positioned.

Defining similar polygons

As you see in Figure 8-1, quadrilateral *WXYZ* is the same shape as quadrilateral *ABCD,* but it's ten times larger (though not drawn to scale). These quadrilaterals are therefore similar.

Similar polygons: For two polygons to be similar, both of the following must be true:

▸ Corresponding angles are congruent.

▸ Corresponding sides are proportional.

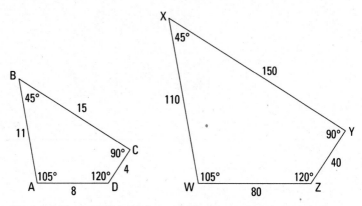

Figure 8-1: These quadrilaterals are *similar* because they're exactly the same shape; note that their angles are congruent.

To fully understand this definition, you have to know what *corresponding angles* and *corresponding sides* mean. Here's the lowdown on *corresponding*. In Figure 8-1, if you expand *ABCD* to the same size as *WXYZ* and slide it to the right, it'd stack perfectly on top of *WXYZ. A* would stack on *W, B* on *X, C* on *Y,* and *D* on *Z.* These vertices are thus *corresponding.* And therefore, you say that ∠*A* corresponds to ∠*W,* ∠*B* corresponds to ∠*X,* and so on. Also, side $\overline{AB}$ corresponds to side $\overline{WX}$, $\overline{BC}$ to $\overline{XY}$, and so on.

When you name similar polygons, pay attention to how the vertices pair up. For the quadrilaterals in Figure 8-1, you write that *ABCD ~ WXYZ* (the squiggle symbol means *is similar to*) because *A* and *W* (the first letters) are corresponding vertices, *B* and *X* (the second letters) are corresponding, and so on. You can also write *BCDA ~ XYZW* (because corresponding vertices pair up) but *not ABCD ~ YZWX.*

Now I'll use quadrilaterals *ABCD* and *WXYZ* to explore the definition of similar polygons in greater depth:

> ✔ **Corresponding angles are congruent.** You can see that ∠*A* and ∠*W* are both 105° and thus congruent, ∠*B* ≅ ∠*X,* and so on. When you blow up or shrink a figure, the angles don't change.

✔ **Corresponding sides are proportional.** The ratios of corresponding sides are equal, like this:

$$\frac{\text{left side}_{WXYZ}}{\text{left side}_{ABCD}} = \frac{\text{top}_{WXYZ}}{\text{top}_{ABCD}} = \frac{\text{right side}_{WXYZ}}{\text{right side}_{ABCD}} = \frac{\text{base}_{WXYZ}}{\text{base}_{ABCD}}$$

$$\frac{WX}{AB} = \frac{XY}{BC} = \frac{YZ}{CD} = \frac{ZW}{DA}$$

$$\frac{110}{11} = \frac{150}{15} = \frac{40}{4} = \frac{80}{8} = 10$$

Each ratio equals 10, the expansion factor. (If the ratios were flipped upside down — which is equally valid — each would equal $\frac{1}{10}$, the shrink factor.) And not only do these ratios all equal 10, but the ratio of the perimeters of *ABCD* and *WXYZ* also equals 10.

How similar figures line up

Two similar figures can be positioned so that they either line up or don't line up. You can see that figures *ABCD* and *WXYZ* in Figure 8-1 are positioned in the same way in the sense that if you were to blow up *ABCD* to the size of *WXYZ* and then slide *ABCD* over, it'd match up perfectly with *WXYZ*. Now check out Figure 8-2, which shows *ABCD* again with another similar quadrilateral. You can easily see that, unlike the quadrilaterals in Figure 8-1, *ABCD* and *PQRS* are *not* positioned in the same way.

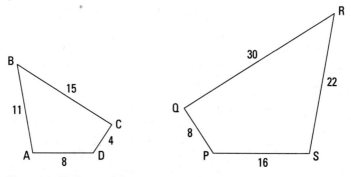

Figure 8-2: Similar quadrilaterals that aren't lined up.

In the preceding section, you see how to set up a proportion for similar figures using the positions of their sides, which I've labeled *left side, right side, top,* and *base* — for example, one valid proportion is $\frac{\text{left side}}{\text{left side}} = \frac{\text{top}}{\text{top}}$. This is a good way to think about how proportions work with similar figures, but this works only if the figures are drawn like *ABCD* and *WXYZ* are. When similar figures are drawn facing different ways, as in Figure 8-2, the left side doesn't necessarily correspond to the left side, and so on, and you have to take greater care that you're pairing up the proper vertices and sides.

Quadrilaterals *ABCD* and *PQRS* are similar, but you *can't* say that *ABCD ~ PQRS* because the vertices don't pair up in this order. Ignoring its size, *PQRS* is the mirror image of *ABCD*. If you flip *PQRS* over in the left-right direction, you get the image in Figure 8-3.

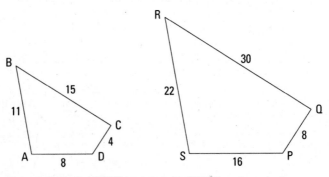

Figure 8-3: Flipping *PQRS* over to make *SRQP* lines it up nicely with *ABCD* — pure poetry!

Now it's easier to see how the vertices pair up. *A* corresponds to *S, B* with *R,* and so on, so you write the similarity like this: *ABCD ~ SRQP.*

Align similar polygons. If you get a problem with a diagram of similar polygons that aren't lined up, consider redrawing one of them so that they're both positioned in the same way. This may make the problem easier to solve.

Solving a similarity problem

Enough of this general stuff — let's see these ideas in action:

Given: *ROTFL ~ SUBAG*

Perimeter of

ROTFL is 52

Find: 1. The lengths of

$\overline{AG}$ and $\overline{GS}$

2. The perimeter

of *SUBAG*

3. The measures of

$\angle S$, $\angle G$, and $\angle A$

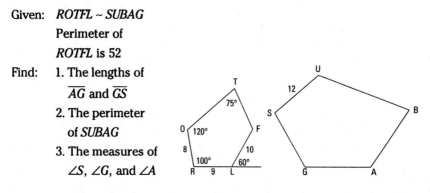

You can see that the *ROTFL* and *SUBAG* aren't positioned the same way just by looking at the figure (and noting that their first letters, *R* and *S*, aren't in the same place). So you need to make sure you pair up their vertices correctly, but that's a snap because the letters in *ROTFL ~ SUBAG* show you what corresponds to what. *R* corresponds to *S*, *O* corresponds to *U*, and so on. (By the way, do you see what you'd have to do to line up *SUBAG* with *ROTFL*? *SUBAG* has sort of been tipped over to the right, so you'd have to rotate it counterclockwise a bit and stand it up on base $\overline{GS}$. You may want to redraw *SUBAG* like that, which can really help you see how all the parts of the two pentagons correspond.)

1. Find the lengths of $\overline{AG}$ and $\overline{GS}$.

The order of the vertices in *ROTFL ~ SUBAG* tells you that $\overline{SU}$ corresponds to $\overline{RO}$ and that $\overline{AG}$ corresponds to $\overline{FL}$; thus, you can set up the following proportion to find missing length $\overline{AG}$:

$$\frac{AG}{FL} = \frac{SU}{RO}$$

$$\frac{AG}{10} = \frac{12}{8}$$

$$\frac{AG}{10} = 1.5 \text{ (or you could have cross-multiplied)}$$

$$AG = 15$$

This method of setting up a proportion and solving for the unknown length is the standard way of solving this type of problem. It's often useful, and you should know how to do it.

But another method can come in handy. Here's how to use it to find $\overline{GS}$: Divide the lengths of two known sides of the figures like this: $\frac{SU}{RO} = \frac{12}{8}$, which equals 1.5. That answer tells you that all the sides of *SUBAG* (and its perimeter) are 1.5 times as long as their counterparts in *ROTFL*. The order of the vertices in *ROTFL ~ SUBAG* tells you that $\overline{GS}$ corresponds to $\overline{LR}$; thus, $\overline{GS}$ is 1.5 times as long as $\overline{LR}$:

$$GS = 1.5 \cdot LR$$
$$= 1.5 \cdot 9$$
$$= 13.5$$

2. **Find the perimeter of *SUBAG*.**

   The method I just introduced tells you that

   $$\text{Perimeter}_{SUBAG} = 1.5 \cdot \text{Perimeter}_{ROTFL}$$
   $$= 1.5 \cdot 52$$
   $$= 78$$

3. **Find the measures of $\angle S$, $\angle G$, and $\angle A$.**

   S corresponds to *R*, *G* corresponds to *L*, and *A* corresponds to *F*, so

   - Angle *S* is the same as $\angle R$, or 100°.
   - Angle *G* is the same as $\angle RLF$, which is 120° (the supplement of the 60° angle).

   To get $\angle A$, you first have to find $\angle F$ with the sum-of-angles formula:

   $$\text{Sum of angles}_{PentagonROTFL} = (n-2)180$$
   $$= (5-2)180$$
   $$= 540°$$

   Because the other four angles of *ROTFL* (clockwise from *L*) add up to 120° + 100° + 120° + 75° = 415°, $\angle F$, and therefore $\angle A$, must equal 540° – 415°, or 125°.

Proving Triangles Similar

Chapter 5 explains five ways to prove triangles congruent: SSS, SAS, ASA, AAS, and HLR. Here, I show you the three ways to prove triangles *similar:* AA, SSS~, and SAS~.

Use the following methods to prove triangles similar:

✔ **AA:** If two angles of one triangle are congruent to two angles of another triangle, then the triangles are similar.

✔ **SSS~:** If the ratios of the three pairs of corresponding sides of two triangles are equal, then the triangles are similar.

✔ **SAS~:** If the ratios of two pairs of corresponding sides of two triangles are equal and the included angles are congruent, then the triangles are similar.

Tackling an AA proof

The AA method is the most frequently used and is therefore the most important. Luckily, it's also the easiest of the three methods to use. Give it a whirl with the following proof:

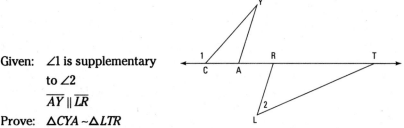

Given: ∠1 is supplementary
         to ∠2
         $\overline{AY} \parallel \overline{LR}$
Prove: △*CYA* ~ △*LTR*

Whenever you see parallel lines in a similar-triangle problem, look for ways to use the parallel-line theorems from Chapter 6 to get congruent angles.

Here's a possible game plan (this hypothetical thought process assumes that you don't know that this is an AA proof from the title of this section): The first given is about angles, and the second given is about parallel lines, which will probably tell you something about congruent angles. Therefore, this proof is almost certainly an AA proof. So all you have to do is think about the givens and figure out which two pairs of angles you can prove congruent to use for AA. Duck soup.

Take a look at how the proof plays out:

Statements	Reasons
1) ∠1 is supplementary to ∠2	1) Given.
2) ∠1 is supplementary to ∠YCA	2) Two angles that form a straight angle (assumed from diagram) are supplementary.
3) ∠YCA ≅ ∠2	3) Supplements of the same angle are congruent.
4) $\overline{AY} \parallel \overline{LR}$	4) Given.
5) ∠CAY ≅ ∠LRT	5) Alternate exterior angles are congruent (using parallel segments $\overline{AY}$ and $\overline{LR}$ and transversal $\overleftrightarrow{CT}$).
6) △CYA ≅ △LTR	6) AA. (If two angles of one triangle are congruent to two angles of another triangle, then the triangles are similar; lines 3 and 5.)

Using SSS~

The upcoming SSS~ proof incorporates the Midline Theorem.

The Midline Theorem: A segment joining the midpoints of two sides of a triangle — which is called a *midline* — is

✔ One-half the length of the third side, and

✔ Parallel to the third side.

Check out this theorem in action with an SSS~ proof:

Given: *A, W,* and *Y* are the
         midpoints of $\overline{KN}$, $\overline{KE}$, and
         $\overline{NE}$, respectively

Prove: In a paragraph proof, show
         that $\triangle WAY \sim \triangle NEK$ using
         1. The first part of the
            Midline Theorem
         2. The second part of the
            Midline Theorem

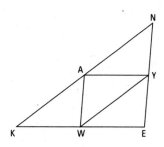

1. Use the first part of the Midline Theorem to prove that $\triangle WAY \sim \triangle NEK$.

The first part of the Midline Theorem says that a segment connecting the midpoints of two sides of a triangle is half the length of the third side. You have three such segments: $\overline{AY}$ is half the length of $\overline{KE}$, $\overline{WY}$ is half the length of $\overline{KN}$, and $\overline{AW}$ is half the length of $\overline{NE}$. That gives you the proportionality you need: $\frac{AY}{KE} = \frac{WY}{KN} = \frac{AW}{NE} = \frac{1}{2}$. Thus, the triangles are similar by SSS~.

2. Use part two of the Midline Theorem to prove that $\triangle WAY \sim \triangle NEK$.

The second part of the Midline Theorem tells you that a segment connecting the midpoints of two sides of a triangle is parallel to the third side. You have three segments like this, $\overline{AY}$, $\overline{WY}$, and $\overline{AW}$, each of which is parallel to a side of $\triangle NEK$. The pairs of parallel segments should make you think about using the parallel-line theorems, which could give you the congruent angles you need to prove the triangles similar with AA.

Look at parallel segments $\overline{AY}$ and $\overline{KE}$, with transversal $\overline{NE}$. You can see that $\angle E$ is congruent to $\angle AYN$ because corresponding angles (the parallel-line meaning of *corresponding angles*) are congruent.

Now look at parallel segments $\overline{AW}$ and $\overline{NE}$, with transversal $\overline{AY}$. Angle AYN is congruent to $\angle WAY$ because they're alternate interior angles. So by the Transitive Property, $\angle E \cong \angle WAY$.

With identical reasoning, you show that $\angle K \cong \angle WYA$ or that $\angle N \cong \angle AWY$. The end. The triangles are similar by AA.

An SAS~ proof

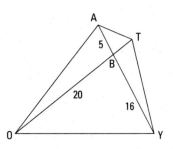

Given: $\triangle BOA \sim \triangle BYT$

Prove: $\triangle BAT \sim \triangle BOY$

(paragraph proof)

Game Plan: Your thinking might go like this. You have one pair of congruent angles, the vertical angles $\angle ABT$ and $\angle OBY$. But it doesn't look like you can get another pair of congruent angles, so the AA approach is out. What other method can you try? You're given side lengths in the figure, so the combination of angles and sides should make you think of SAS~. To prove $\triangle BAT \sim \triangle BOY$ with SAS~, you need to find the length of $\overline{BT}$ so you can show that $\overline{BA}$ and $\overline{BT}$ (the sides that make up $\angle ABT$) are proportional to $\overline{BO}$ and $\overline{BY}$ (the sides that make up $\angle OBY$). To find $\overline{BT}$, you can use the similarity in the given.

So you begin solving the problem by figuring out the length of $\overline{BT}$. $\triangle BOA \sim \triangle BYT$, so — paying attention to the order of the letters — you see that $\overline{BO}$ corresponds to $\overline{BY}$ and that $\overline{BA}$ corresponds to $\overline{BT}$. Thus, you can set up this proportion:

$$\frac{BO}{BY} = \frac{BA}{BT}$$

$$\frac{20}{16} = \frac{5}{BT}$$

$$20 \cdot BT = 16 \cdot 5$$

$$BT = 4$$

Now, to prove $\triangle BAT \sim \triangle BOY$ with SAS~, you use the congruent vertical angles and then check that the following proportion works:

$$\frac{BA}{BO} \overset{?}{=} \frac{BT}{BY}$$

$$\frac{5}{20} \overset{?}{=} \frac{4}{16}$$

This checks. You're done. (By the way, these fractions both reduce to $\frac{1}{4}$, so $\triangle BAT$ is $\frac{1}{4}$ as big as $\triangle BOY$.)

Splitting Right Triangles with the Altitude-on-Hypotenuse Theorem

In a right triangle, the altitude that's perpendicular to the hypotenuse has a special property: It creates two smaller right triangles that are both similar to the original right triangle.

Altitude-on-Hypotenuse Theorem: If an altitude is drawn to the hypotenuse of a right triangle as shown in Figure 8-4, then

✔ The two triangles formed are similar to the given triangle and to each other:

$$\triangle ACB \sim \triangle ADC \sim \triangle CDB,$$

✔ $h^2 = xy$, and

✔ $a^2 = yc$ and $b^2 = xc$.

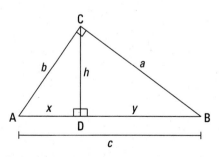

Figure 8-4: Three similar right triangles: small, medium, and large.

Use Figure 8-5 to answer the following questions.

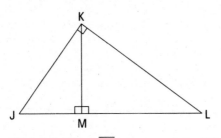

Figure 8-5: Altitude $\overline{KM}$ lets you apply the Altitude-on-Hypotenuse Theorem.

If $ML = 16$ and $JK = 15$, what's $JM?$

Set JM equal to x; then use part three of the theorem.

$$(JK)^2 = (JM)(JL)$$
$$15^2 = x(x+16)$$
$$225 = x^2 + 16x$$
$$x^2 + 16x - 225 = 0$$
$$(x-9)(x+25) = 0$$
$$x - 9 = 0 \quad \text{or} \quad x + 25 = 0$$
$$x = 9 \quad \text{or} \quad x = -25$$

You know that a length can't be –25, so $JM = 9$.

When doing a problem involving an altitude-on-hypotenuse diagram, don't assume that you must use the second or third part of the Altitude-on-Hypotenuse Theorem. Sometimes, the easiest way to solve the problem is with the Pythagorean Theorem. And at other times, you can use ordinary similar-triangle proportions to solve the problem.

The next question illustrates this tip: If $JK = 9$ and $KL = 12$, how long is $KM?$

First get *JL* with the Pythagorean Theorem or by noticing that you have a triangle in the 3 : 4 : 5 family — namely a 9-12-15 triangle. So *JL* = 15. Now you could finish with the Altitude-on-Hypotenuse Theorem, but that approach is a bit complicated. Instead, just use an ordinary similar-triangle proportion:

$$\frac{\text{long leg}_{\triangle JKM}}{\text{long leg}_{\triangle JLK}} = \frac{\text{hypotenuse}_{\triangle JKM}}{\text{hypotenuse}_{\triangle JLK}}$$

$$\frac{KM}{12} = \frac{9}{15}$$

$$15 \cdot KM = 108$$

$$KM = 7.2$$

More Proportionality Theorems

In this section, you get two more theorems that, like similar polygons, involve proportions.

The Side-Splitter Theorem

The Side-Splitter Theorem isn't really necessary because the problems in which you use it involve similar triangles, so you can solve them with the ordinary similar-triangle proportions from earlier in this chapter. The Side-Splitter Theorem just gives you an alternative, shortcut solution method.

Side-Splitter Theorem: If a line is parallel to a side of a triangle and it intersects the other two sides, it divides those sides proportionally.

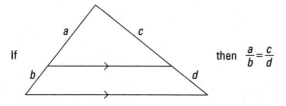

If ... then $\dfrac{a}{b} = \dfrac{c}{d}$

Figure 8-6: A line parallel to a side cuts the other two sides proportionally.

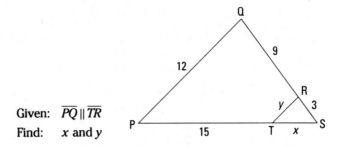

Given: $\overline{PQ} \parallel \overline{TR}$

Find: x and y

Because $\overline{PQ} \parallel \overline{TR}$, you use the Side-Splitter Theorem to get x:

$$\frac{x}{15} = \frac{3}{9}$$
$$9x = 45$$
$$x = 5$$

Now find y: First, don't fall for the trap and conclude that $y = 4$. This is a doubly sneaky trap that I'm especially proud of. Side y looks like it should equal 4 for two reasons: First, you could jump to the erroneous conclusion that $\triangle TRS$ is a 3-4-5 right triangle. But nothing tells you that $\angle TRS$ is a right angle, so you can't conclude that.

Second, when you see the ratios of 9 : 3 (along $\overline{QS}$) and 15 : 5 (along $\overline{PS}$, after solving for x), both of which reduce to 3 : 1, it looks like PQ and y should be in the same 3 : 1 ratio. That would make $PQ : y$ a 12 : 4 ratio, which again leads to the wrong answer that y is 4. The answer comes out wrong because this thought process amounts to using the Side-Splitter Theorem for the sides that aren't split — which you aren't allowed to do.

Don't use the Side-Splitter Theorem on sides that aren't split. You can use the Side-Splitter Theorem *only* for the four segments on the split sides of the triangle. Do *not* use it for the parallel sides, which are in a different ratio. For the parallel sides, use similar-triangle proportions. (Whenever a triangle is divided by a line parallel to one of its sides, the smaller triangle created is similar to the original, larger triangle.)

So finally, the correct way to get y is to use an ordinary similar-triangle proportion. The triangles in this problem are positioned the same way, so you can write the following:

$$\frac{\text{left side}_{\triangle TRS}}{\text{left side}_{\triangle PQS}} = \frac{\text{base}_{\triangle TRS}}{\text{base}_{\triangle PQS}}$$

$$\frac{y}{12} = \frac{5}{20}$$

$$20y = 60$$

$$y = 3$$

The Angle-Bisector Theorem

In this final section, you get another theorem involving a proportion; but unlike everything else in this chapter, this theorem has nothing to do with similarity.

Angle-Bisector Theorem: If a ray bisects an angle of a triangle, then it divides the opposite side into segments that are proportional to the other two sides. See Figure 8-7.

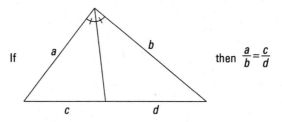

Figure 8-7: Because the angle is bisected, segments c and d are proportional to sides a and b.

Don't forget the Angle-Bisector Theorem. (For some reason, students often do forget this theorem.) Whenever you see a triangle with one of its angles bisected, consider using the theorem.

How about an angle-bisector problem? Why? Oh, just *BCUZ*.

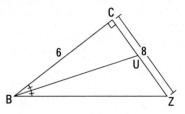

Given: Diagram as shown

Find: *BZ*, *CU*, *UZ*, and *BU*

You get *BZ* with the Pythagorean Theorem ($6^2 + 8^2 = c^2$) or by noticing that $\triangle BCZ$ is in the 3 : 4 : 5 family. It's a 6-8-10 triangle, so *BZ* is 10.

Next, set *CU* equal to x; this makes *UZ* equal to $8 - x$. Set up the angle-bisector proportion and solve for x:

$$\frac{6}{10} = \frac{x}{8-x}$$
$$48 - 6x = 10x$$
$$48 = 16x$$
$$3 = x$$

So *CU* is 3 and *UZ* is 5.

The Pythagorean Theorem then gives you *BU*:

$$(BU)^2 = 6^2 + 3^2$$
$$(BU)^2 = 45$$
$$BU = \sqrt{45} = 3\sqrt{5} \approx 6.7$$

Chapter 9

Circle Basics

In This Chapter

▶ Defining radii, chords, and arcs

▶ Theorizing about the angle-arc theorems

▶ Practicing products with the power theorems

*I*n a sense, the circle is the simplest of all shapes: no corners, no irregularities, the same simple shape no matter how you turn it. On the other hand, that simple curve involves the number pi ($\pi = 3.14159\ldots$), and nothing's simple about that. It goes on forever with no repeating pattern of digits. Despite the fact that mathematicians have been studying the circle and the number π for over 2,000 years, many unsolved mysteries about them remain. Let's get started looking at this fascinating shape.

Radii, Chords, and Diameters

Why don't we begin with the three main types of line segments inside a circle: radii, chords, and diameters:

✔ **Radius:** A circle's radius is the distance from its center to a point on the circle. In addition to being a measure of distance, a radius is also a segment that goes from a circle's center to a point on the circle.

✔ **Chord:** A segment that connects two points on a circle.

✔ **Diameter:** A chord that passes through a circle's center. A circle's diameter is twice as long as its radius.

Five circle theorems

I hope you have some available space on your mental hard drive for more theorems. (If not, maybe you can free up some room by deleting a few not-so-useful facts such as the date of the Battle of Hastings: A.D. 1066.)

✔ **Radii size:** All radii of a circle are congruent.

✔ **Perpendicularity and bisected chords:**

- If a radius is perpendicular to a chord, then it bisects the chord.

- If a radius bisects a chord (that isn't a diameter), then it's perpendicular to the chord.

✔ **Distance and chord size:**

- If two chords of a circle are equidistant from the center of the circle, then they're congruent.

- If two chords of a circle are congruent, then they're equidistant from its center.

Using extra radii

In real estate, the three most important factors are *location, location, location*. With circles, it's *radii, radii, radii*. In circle problems, you'll often need to add radii and partial radii to create right triangles or isosceles triangles that you can then use to solve the problem.

✔ **Draw additional radii on the figure.** You should draw radii to points where something else intersects or touches the circle, as opposed to just any old point on the circle.

✔ **Open your eyes and notice all the radii, including new ones you've drawn, and mark them all congruent.** For some odd reason, people often fail to notice all the radii in a problem or fail to note that they're congruent.

✔ **Draw in the segment (part of a radius) that goes from the center of a circle to a chord and that's perpendicular to the chord.** This segment bisects the chord.

Let's do a problem: Find the area of inscribed quadrilateral *GHJK* shown on the left. The circle has a radius of 2.

The tip above gives you two hints for this problem. The first is to draw in the four radii to the four vertices of the quadrilateral as shown in the figure on the right.

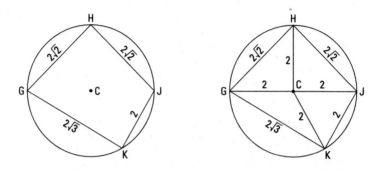

Now you simply need to find the area of the individual triangles. You can see that △JKC is equilateral, so you can use the equilateral triangle formula for this one:

$$\text{Area} = \frac{s^2\sqrt{3}}{4} = \frac{2^2\sqrt{3}}{4} = \sqrt{3} \text{ units}^2$$

And if you're on the ball, you should recognize triangles *GHC* and *HJC*. Their sides are in the ratio of $2 : 2 : 2\sqrt{2}$, which reduces to $1 : 1 : \sqrt{2}$; thus, they're 45°- 45°- 90° *right* triangles. The two legs of a right triangle can be used for its base and height, so getting their areas is a snap. For each triangle,

$$\text{Area} = \frac{1}{2}bh = \frac{1}{2} \cdot 2 \cdot 2 = 2 \text{ units}^2$$

Another hint from the tip helps you with △KGC. Draw its altitude (a partial radius) from C to $\overline{GK}$. This radius is perpendicular to $\overline{GK}$ and thus bisects $\overline{GK}$ into two segments of length $\sqrt{3}$. You've divided △KGC into two right triangles; each has a hypotenuse of 2 and a leg of $\sqrt{3}$, so the other leg (the altitude) is 1 (by the Pythagorean Theorem or by recognizing that these are 30°- 60°- 90° triangles whose sides are in the ratio of $1 : \sqrt{3} : 2$). So △KGC has an altitude of 1 and a base of $2\sqrt{3}$. Just use the regular area formula again:

$$\text{Area} = \frac{1}{2}bh = 2\sqrt{3} \cdot 1 = \sqrt{3} \text{ units}^2$$

Now just add 'em up:

$$\text{Area}_{GHJK} = \text{area}_{\triangle JKC} + \text{area}_{\triangle GHC} + \text{area}_{\triangle HJC} + \text{area}_{\triangle KGC}$$
$$= \sqrt{3} \quad + \quad 2 \quad + \quad 2 \quad + \quad \sqrt{3}$$
$$= 4 + 2\sqrt{3} \approx 7.46 \text{ units}^2$$

Arcs and Central Angles

This section covers arcs and central angles. Big surprise!

✔ **Arc:** An arc is simply a curved piece of a circle. Any two points on a circle divide the circle into two arcs: a *minor arc* (the smaller piece) and a *major arc* (the larger) — unless the points are the endpoints of a diameter, in which case both arcs are semicircles. Figure 9-1 shows minor arc $\overarc{AB}$ (a 60° arc) and major arc $\overarc{ACB}$ (a 300° arc).

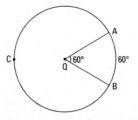

Figure 9-1: A 60° central angle cuts out a 60° arc.

✔ **Central angle:** A central angle is an angle whose vertex is at the center of a circle. The sides of a central angle are radii that hit the circle at the opposite ends of an arc — or as mathematicians say, the angle *intercepts* the arc.

The measure of an arc is the same as the degree measure of the central angle that intercepts it. The figure shows central angle $\angle AQB$, which, like $\overarc{AB}$, measures 60°.

Tangents

This section covers — can you guess? — tangent lines. A line is *tangent* to a circle if it touches it at one and only one point.

Radius-tangent perpendicularity: If a line is tangent to a circle, then it is perpendicular to the radius drawn to the point of tangency.

Don't neglect to check circle problems for tangent lines and the right angles that occur at points of tangency. You may have to draw in one or more radii to points of tangency to create the right angles. The right angles often become parts of right triangles (or sometimes rectangles).

Here's an example problem: Find the radius of circle C and the length of $\overline{DE}$ in the following figure.

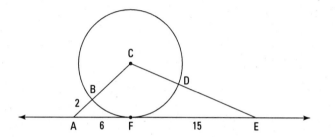

When you see a circle problem, you should be saying to yourself: *radii, radii, radii!* So draw in radius $\overline{CF}$, which, according to the above theorem, is perpendicular to $\overleftrightarrow{AE}$. Set it equal to x, which gives $\overline{CB}$ a length of x as well. You now have right triangle $\triangle CFA$, so use the Pythagorean Theorem to find x:

$$x^2 + 6^2 = (x+2)^2$$
$$x^2 + 36 = x^2 + 4x + 4$$
$$32 = 4x$$
$$8 = x$$

So the radius is 8. Then you can see that $\triangle CFE$ is an 8-15-17 triangle, so CE is 17. (Of course, you can also get CE with the Pythagorean Theorem.) CD is 8 (and it's the third radius in this problem; does *radii, radii, radii* ring a bell?). Therefore, DE is $17 - 8$, or 9. That does it.

The Pizza Slice Formulas

In this section, you begin with the formulas for a circle's area and circumference. Then you use those formulas to compute lengths, perimeters, and areas of various parts of a circle.

Circumference and area of a circle:

> ✔ Circumference = $2\pi r$ (or πd)
>
> ✔ Area$_{Circle}$ = πr^2

Determining arc length

Before getting to the arc length formula, I want to mention a potential source of confusion about arcs. Earlier in this chapter, the *measure of an arc* is defined as the degree measure of the central angle that intercepts the arc. But in this section, I go over the *length of an arc*. In this context, *length* means the same commonsense thing length always means — you know, like the length of a piece of string. (With an arc, of course, it'd be a curved piece of string.) In a nutshell, the *measure* of an arc is the degree size of its central angle; the *length* of an arc is the regular length along the arc.

A circle is 360° all the way around; therefore, if you divide an arc's degree measure by 360°, you find the fraction of the circle's circumference that the arc makes up. Then, if you multiply the length of the circumference by that fraction, you get the length along the arc.

Arc length: The length of an *arc* is equal to the circumference of the circle ($2\pi r$) times the fraction of the circle represented by the arc's measure (note that the degree measure of an arc is written like $m\overset{\frown}{AB}$):

$$\text{Length}_{\overset{\frown}{AB}} = \left(\frac{m\overset{\frown}{AB}}{360} \right)(2\pi r)$$

See Figure 9-2 for an example.

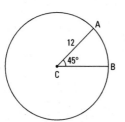

Figure 9-2: Arc $\widehat{AB}$ is $\frac{1}{8}$ of the circle's circumference.

Check out the calculations for $\widehat{AB}$. Its degree measure is 45° and the radius of the circle is 12, so here's the math for its length:

$$\text{Length}_{\widehat{AB}} = \left(\frac{m\widehat{AB}}{360}\right)(2\pi r)$$

$$= \frac{45}{360} \cdot 2 \cdot \pi \cdot 12$$

$$= \frac{1}{8} \cdot 24\pi = 3\pi \approx 9.42 \text{ units}$$

As you can see, because 45° is $\frac{1}{8}$ of 360°, the length of arc $\widehat{AB}$ is $\frac{1}{8}$ of the circumference of 24π.

Sector and segment area

In this section, I cover sector and segment area.

- ✔ **Sector:** A region bounded by two radii and an arc of a circle. (Plain English: the shape of a piece of pizza.)
- ✔ **Segment of a circle:** A region bounded by a chord and an arc of a circle.

Just as an arc is a fraction of a circle's circumference, a sector is a fraction of a circle's area; so computing the area of a sector works exactly like the arc-length formula.

Area of a sector: The area of a sector (such as sector *PQR* in Figure 9-3) is equal to the area of the circle (πr^2) times the fraction of the circle represented by the sector:

$$\text{Area}_{\text{Sector}PQR} = \left(\frac{m\widehat{PR}}{360}\right)(\pi r^2)$$

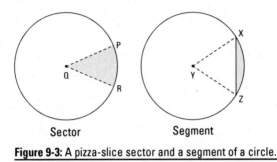

Sector   Segment

Figure 9-3: A pizza-slice sector and a segment of a circle.

Look back to Figure 9-2. When you do the math with the above formula, you'll see that because 45° is $\frac{1}{8}$ of 360°, the area of sector ACB is $\frac{1}{8}$ of the area of the circle (just like the length of $\overarc{AB}$ is $\frac{1}{8}$ of the circle's circumference).

Area of a segment: To compute the area of a segment like the one in Figure 9-3, just subtract the area of the triangle from the area of the sector (by the way, there's no technical way to name segments, but let's call this one *circle segment XZ*):

$$\text{Area}_{\text{Circle Segment}XZ} = \text{area}_{\text{sector}XYZ} - \text{area}_{\triangle XYZ}$$

We just covered how to compute the area of a sector. To get the triangle's area, you draw an altitude that goes from the circle's center to the chord that makes up the triangle's base. This altitude is a leg of a right triangle whose hypotenuse is a radius of the circle. You finish with right-triangle ideas such as the Pythagorean Theorem. You see this in the next problem.

Given: Circle D with a radius of 6

Find: 1. Length of arc $\overarc{IK}$

2. Area of sector IDK

3. Area of circle segment IK

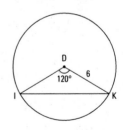

1. Find the length of arc $\overarc{IK}$.

The measure of the arc is 120°, which is a third of 360°, so the length of $\overarc{IK}$ is a third of the circumference.

That's all there is to it. Here's how this looks when you plug it into the formula:

$$\text{Length}_{\overset{\frown}{IK}} = \left(\frac{m\overset{\frown}{IK}}{360} \right)(2\pi r)$$
$$= \frac{120}{360} \cdot 12\pi$$
$$= \frac{1}{3} \cdot 12\pi = 4\pi \approx 12.6 \text{ units}$$

2. Find the area of sector IDK.

Because 120° takes up a third of the degrees in a circle, sector *IDK* occupies a third of the circle's area.

$$\text{Area}_{\text{Sector}IDK} = \left(\frac{m\overset{\frown}{IK}}{360} \right)(\pi r^2)$$
$$= \frac{120}{360} \cdot 36\pi$$
$$= \frac{1}{3} \cdot 36\pi = 12\pi \approx 37.7 \text{ units}^2$$

3. Find the area of circle segment IK.

To get this, you need the area of $\triangle IDK$ so you can subtract it from the area of sector *IDK*. Draw an altitude straight down from *D* to $\overline{IK}$. That creates two 30°- 60°- 90° triangles. The sides of a 30°- 60°- 90° triangle are in the ratio of $x : x\sqrt{3} : 2x$. In this problem, the hypotenuse is 6, so the altitude (the short leg) is half of that, or 3, and the base (the long leg) is $3\sqrt{3}$. $\overline{IK}$ is twice as long as the base of the 30°- 60°- 90° triangle, so it's twice $3\sqrt{3}$, or $6\sqrt{3}$. You finish with the segment area formula:

$$\text{Area}_{\text{Segment}IDK} = \text{area}_{\text{Sector}IDK} - \text{area}_{\triangle IDK}$$
$$= 12\pi - \frac{1}{2}bh \quad \text{(You got the } 12\pi \text{ in part 2)}$$
$$= 12\pi - \frac{1}{2} \cdot 6\sqrt{3} \cdot 3$$
$$= 12\pi - 9\sqrt{3}$$
$$\approx 22.1 \text{ units}^2$$

The Angle-Arc Formulas

This section looks at angles that intersect a circle. The vertices of these angles can lie *inside* the circle, *on* the circle, or *outside* the circle. The formulas in this section tell you how these angles are related to the arcs they intercept.

Angles on a circle

Of the three places an angle's vertex can be in relation to a circle, the angles whose vertices lie *on* a circle are the ones that come up in the most problems and are therefore the most important. These angles come in two flavors:

- **Inscribed angle:** An inscribed angle, like $\angle BCD$ in Figure 9-4a, is an angle whose vertex lies on a circle and whose sides are two chords of the circle.

- **Tangent-chord angle:** A tangent-chord angle, like $\angle JKL$ in Figure 9-4b, is an angle whose vertex lies on a circle and whose sides are a tangent and a chord of the circle.

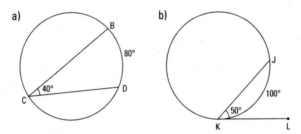

Figure 9-4: Angles with vertices on a circle.

 Measure of an angle on a circle: The measure of an inscribed angle or a tangent-chord angle is *one-half* the measure of its intercepted arc.

For example, in Figure 9-4, $\angle BCD = \frac{1}{2}\left(m\widehat{BD}\right)$ and $\angle JKL = \frac{1}{2}\left(m\widehat{JK}\right)$.

 Make sure you remember the simple idea that an angle on a circle is half the measure of the arc it intercepts. If you forget which is half of which, try this: Draw a quick sketch of a circle with a 90° arc (a quarter of the circle) and an inscribed angle that intercepts the 90° arc. You'll see right away that the angle is less than 90°, telling you that the angle is the thing that's half of the arc, not vice versa.

Angles inside a circle

Measure of an angle inside a circle: The measure of an angle whose vertex is *inside* a circle (a *chord-chord angle*) is one-half the sum of the measures of the arcs intercepted by the angle and its vertical angle. Check out Figure 9-5, which shows you chord-chord angle *SVT*. You find the measure of the angle like this:

$$\angle SVT = \frac{1}{2}\left(m\widehat{ST} + m\widehat{QR}\right)$$

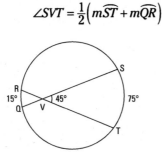

Figure 9-5: Chord-chord angles are inside a circle.

Angles outside a circle

Three varieties of angles fall outside a circle, and all are made up of tangents and secants. You know what a tangent is, and here's the definition of *secant*.

Technically, a *secant* is a line that intersects a circle at two points. But the secants you use in this section are segments that cut through a circle and that have one endpoint outside the circle and one endpoint on the circle.

So here are the three types of angles that are *outside* a circle:

- ✔ **Secant-secant angle:** A secant-secant angle, like ∠*BDF* in Figure 9-6a, is an angle whose vertex lies outside a circle and whose sides are two secants.

- ✔ **Secant-tangent angle:** A secant-tangent angle, like ∠*GJK* in Figure 9-6b, is an angle whose vertex lies outside a circle and whose sides are a secant and a tangent.

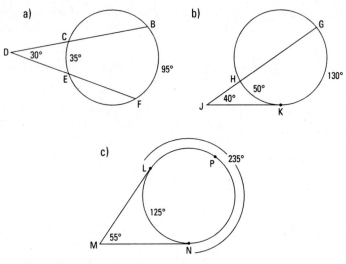

✔ **Tangent-tangent angle:** A tangent-tangent angle, like ∠*LMN* in Figure 9-6c, is an angle whose vertex lies outside a circle and whose sides are two tangents.

Measure of an angle outside a circle: The measure of a secant-secant angle, a secant-tangent angle, or a tangent-tangent angle is one-half the difference of the measures of the intercepted arcs. For example, in Figure 9-6,

$$\angle BDF = \frac{1}{2}\left(m\widehat{BF} - m\widehat{CE}\right)$$

$$\angle GJK = \frac{1}{2}\left(m\widehat{GK} - m\widehat{HK}\right)$$

$$\angle LMN = \frac{1}{2}\left(m\widehat{LPN} - m\widehat{LN}\right)$$

Note that you should subtract the smaller arc from the larger. (If you get a negative answer, you subtracted the wrong way.)

Keeping the formulas straight

In the previous three sections, you see six types of angles made up of chords, secants, and tangents but only three angle-arc formulas. As you can tell from the titles of the sections, to determine which of the three angle-arc formulas you need to use, all you need to pay attention to is where the angle's vertex is: inside, on, or outside the circle. You don't have to worry about whether the two sides of the angle are chords, tangents, secants, or some combination of these things.

The next tip can help you remember which formula goes with which category of angle. First, check out Figure 9-7.

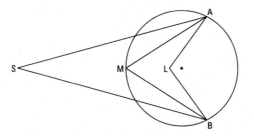

Figure 9-7: As the angle gets farther from the center of the circle, it gets smaller.

You can see that the *small* angle, $\angle S$ (maybe about 35°) is *outside* the circle; the *medium* angle, $\angle M$ (about 70°) is *on* the circle; and the *large* angle, $\angle L$ (roughly 110°) is *inside* the circle. Here's one way to understand why the sizes of the angles go in this order. Say that the sides of $\angle L$ are elastic. Picture grabbing $\angle L$ at its vertex and pulling it to the left (as its ends remain attached to A and B). The farther you pull $\angle L$ to the left, the smaller the angle would get.

Subtracting makes things smaller, and adding makes things larger, right? So here's how to remember which angle-arc formula to use (see Figure 9-7 and Figures 9-4, 9-5, and 9-6):

- ✔ To get the *small* angle, you *subtract:* $\angle S = \frac{1}{2}(\text{arc} - \text{arc})$
- ✔ To get the *medium* angle, you *do nothing:* $\angle M = \frac{1}{2}(\text{arc})$
- ✔ To get the *large* angle, you *add:* $\angle L = \frac{1}{2}(\text{arc} + \text{arc})$

(***Note:*** For the third bullet above, you have to look at Figure 9-5 to see both arcs; adding the second arc for angle L in Figure 9-7 would have made the figure look too messy.)

The Power Theorems

Like the preceding sections, this section takes a look at what happens when angles and circles intersect. But this time, instead of analyzing the size of angles and arcs, you analyze the lengths of the segments that make up the angles.

The Chord-Chord Theorem

The Chord-Chord Power Theorem was brilliantly named for the fact that the theorem uses a chord and . . . . another chord!

Chord-Chord Power Theorem: If two chords of a circle intersect, then the product of the measures of the parts of one chord is equal to the product of the measures of the parts of the other chord. (Whew, what a mouthful!)

Try out your power-theorem skills on this problem:

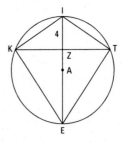

Given: Circle *A* has a radius of 6.5

  KITE is a kite

  IZ = 4

Find: The area of *KITE*

To get the kite's area, you need the lengths of its diagonals. The diagonals are two chords that cross each other, so you should consider using the Chord-Chord Power Theorem.

But first, note that diagonal $\overline{IE}$ is the circle's diameter. Circle *A* has a radius of 6.5, so its diameter is twice as long, or 13, and thus that's the length of diagonal $\overline{IE}$. Then you see that *ZE* must be 13 – 4, or 9. Now you have two of the lengths, *IZ* = 4 and *ZE* = 9, for the segments you use in the theorem:

$$(KZ)(ZT) = (IZ)(ZE)$$

Because *KITE* is a kite, diagonal $\overline{IE}$ bisects diagonal $\overline{KT}$. Thus, $\overline{KZ} \cong \overline{ZT}$, so you can set them both equal to *x*. Plug everything into the equation:

$$x \cdot x = 4 \cdot 9$$
$$x^2 = 36$$
$$x = 6 \text{ or } -6$$

You can obviously reject –6 as a length, so x is 6. KZ and ZT are thus both 6, and diagonal $\overline{KT}$ is therefore 12. You've already figured out that the length of the other diagonal is 13, so now you finish with the kite area formula:

$$\text{Area}_{KITE} = \frac{1}{2}d_1d_2$$
$$= \frac{1}{2} \cdot 12 \cdot 13$$
$$= 78 \text{ units}^2$$

The Tangent-Secant Theorem

Now let's move on to the Tangent-Secant Power Theorem — another awe-inspiring example of creative nomenclature.

Tangent-Secant Power Theorem: If a tangent and a secant are drawn from an external point to a circle, then the square of the measure of the tangent is equal to the product of the measures of the secant's external part and the entire secant. (Another mouthful!)

For example, in Figure 9-8, $8^2 = 4(4 + 12)$

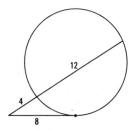

Figure 9-8: The Tangent-Secant Power Theorem: (tangent)^2 = (outside) · (whole).

The Secant-Secant Theorem

Last but not least, I give you the Secant-Secant Power Theorem. Are you sitting down? This theorem involves two secants! (If you're trying to come up with a creative name for your child like Dweezil or Moon Unit, talk to Frank Zappa, not the guy who named the power theorems.)

Secant-Secant Power Theorem: If two secants are drawn from an external point to a circle, then the product of the measures of one secant's external part and that entire secant is equal to the product of the measures of the other secant's external part and that entire secant.

For instance, in Figure 9-9, 4(4 + 2) = 3(3 +5)

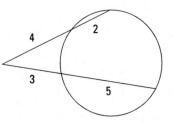

Figure 9-9: The Secant-Secant Power Theorem:
(outside) · (whole) = (outside) · (whole).

Condensing the power theorems into a single idea

All three of the power theorems involve an equation with a product of two lengths (or one length squared) that equals another product of lengths. And each length is a distance from the vertex of an angle to the edge of the circle. Thus, all three theorems use the same scheme:

(vertex to circle) · (vertex to circle) =
(vertex to circle) · (vertex to circle)

This unifying scheme can help you remember all three of the theorems I discuss in the preceding sections. And it'll help you avoid the common mistake of multiplying the external part of a secant by its internal part (instead of correctly multiplying the external part by the entire secant) when you're using the Tangent-Secant Theorem or the Secant-Secant Power Theorem.

Chapter 10

3-D Geometry

*I*n this chapter, you study cones, spheres, prisms, and other solids of varying shapes, focusing on their two most fundamental characteristics, namely *volume* and *surface area*.

Flat-Top Figures

Flat-top figures (that's what I call them, anyway) are solids with two congruent, parallel bases (the top and bottom). A *prism* — your standard cereal box is one example — has polygon-shaped bases, and a *cylinder* — like your standard soup can — has round bases. But despite the different shape of their bases, the same volume and surface area formulas work for both of them because they share the flat-top structure.

▶ **Prism:** A prism is a solid figure with two congruent, parallel, polygonal bases. Its corners are called *vertices,* the segments that connect the vertices are called *edges,* and the flat sides are called *faces.*

▶ **Cylinder:** A cylinder is a solid figure with two congruent, parallel bases that have rounded sides (in other words, the bases are not straight-sided polygons); these bases are connected by a rounded surface.

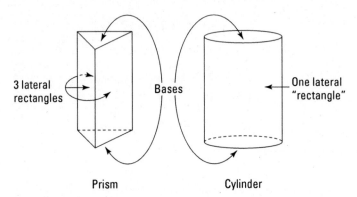

3 lateral rectangles

Bases

One lateral "rectangle"

Prism

Cylinder

Figure 10-1: A prism and a cylinder with their bases and lateral rectangles.

Now that you know what these things are, here are their volume and surface area formulas.

Volume of flat-top figures: The volume of a prism or cylinder is given by the following formula.

$$\text{Vol}_{\text{Flat-top}} = \text{area}_{\text{base}} \cdot \text{height}$$

An ordinary box is a special case of a prism, so you can use the flat-top volume formula for a box, but you probably already know the other way to compute a box's volume: $\text{Vol}_{\text{Box}} = \text{length} \cdot \text{width} \cdot \text{height}$. (Because the length times the width gives you the area of the base, these two methods really amount to the same thing.) To get the volume of a cube, the simplest type of box, you just take the length of one of its edges (sides) and raise it to the third power ($\text{Vol}_{\text{Cube}} = \text{side}^3$).

Surface area of flat-top figures: To find the surface area of a prism or cylinder, use the following formula.

$$\text{SA}_{\text{Flat-top}} = 2 \cdot \text{area}_{\text{base}} + \text{lateral area}_{\text{rectangle(s)}}$$

Because prisms and cylinders have two congruent bases, you simply find the area of one base and double that value; then you add the figure's lateral area. The *lateral area* of a prism or cylinder is the area of the sides of the figure — namely, the area of everything but the figure's bases. Here's how the two figures compare:

✔ **The lateral area of a prism is made up of rectangles.** The bases of a prism can be any shape, but the lateral area is always made up of rectangles. So to get the lateral area, you just find the area of the rectangles using the standard rectangle area formula and then add them up.

✔ **The lateral area of a cylinder is basically one rectangle rolled into a tube shape.** Think of the lateral area of a cylinder as one rectangular paper towel that rolls exactly once around a paper towel roll. The base of this rectangle (you know, the part of the towel that wraps around the bottom of the roll) is the same as the circumference of the cylinder's base. And the height of the paper towel rectangle is the same as the height of the cylinder. So, the lateral area of a cylinder (the area of this rectangle) is $(2\pi r)(h)$, namely $2\pi rh$, or πdh.

Time to take a look at some of these formulas in action.

Given: Prism as shown

$ABCD$ is a square with a diagonal of 8

$\angle EAD$ and $\angle EDA$ are 45° angles

Find: 1. The volume of the prism

2. The surface area of the prism

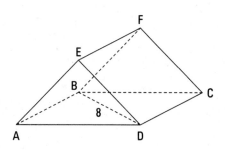

1. Find the volume of the prism.

To use the volume formula, you need the prism's height ($\overline{CD}$) and the area of its base ($\triangle AED$). You've probably noticed that this prism is lying on its side. That's why its height isn't vertical and its base isn't on the bottom.

Get the height first. $ABCD$ is a square, so $\triangle BCD$ (half of the square) is a 45°- 45°- 90° triangle with a hypotenuse of 8. To get the leg of a 45°- 45°- 90° triangle, you divide the hypotenuse by $\sqrt{2}$ (or use the Pythagorean Theorem, noting that $a = b$ in this case). So that gives you $\frac{8}{\sqrt{2}} = 4\sqrt{2}$ for the length of $\overline{CD}$, which, again, is the height of the prism.

And here's how you get the area of $\triangle AED$: First, note that AD, like CD, is $4\sqrt{2}$ (because $ABCD$ is a square). Next, because $\angle EAD$ and $\angle EDA$ are given 45° angles, $\angle AED$ must be 90°; thus, $\triangle AED$ is another 45°- 45°- 90° triangle. Its hypotenuse, $\overline{AD}$, has a length of $4\sqrt{2}$, so its legs ($\overline{AE}$ and $\overline{DE}$) are $\frac{4\sqrt{2}}{\sqrt{2}}$, or 4 units long. The area of a right triangle is given by half the product of its legs (because you can use one leg for the triangle's base and the other for its height), so Area $_{\triangle AED} = \frac{1}{2} \cdot 4 \cdot 4 = 8$. You're all set to finish with the volume formula:

$$\text{Vol}_{\text{Prism}} = \text{area}_{\text{base}} \cdot \text{height}$$
$$= 8 \cdot 4\sqrt{2}$$
$$= 32\sqrt{2}$$
$$\approx 45.3 \text{ units}^3$$

2. Find the surface area of the prism.

Having completed part 1, you have everything you need to compute the surface area. Just plug in the numbers:

$$\text{SA}_{\text{Prism}} = 2 \cdot \text{area}_{\text{base}} \cdot \text{lateral area}_{\text{rectangles}}$$
$$\text{SA} = 2 \cdot 8 + \text{area}_{ABCD} + \text{area}_{AEFB} + \text{area}_{DEFC}$$
$$= 16 + \left(4\sqrt{2}\right)\left(4\sqrt{2}\right) + \left(4\sqrt{2}\right)(4) + \left(4\sqrt{2}\right)(4)$$
$$= 16 + 32 + 16\sqrt{2} + 16\sqrt{2}$$
$$= 48 + 32\sqrt{2}$$
$$\approx 93.3 \text{ units}^2$$

Pointy-Top Figures

What I call *pointy-top figures* are solids with one flat base and . . . a pointy top! The pointy-top solids are the *pyramid* and the *cone*. Even though the pyramid has a polygon-shaped base and the cone has a rounded base, the same volume and surface area formulas work for both.

✔ **Pyramid:** A pyramid is a solid figure with a polygonal base and edges that extend up from the base to meet at a single point.

✔ **Cone:** A cone is a solid figure with a rounded base and a rounded lateral surface that connects the base to a single point.

Volume of pointy-top figures: Here's how to find the volume of a pyramid or cone.

$$\text{Vol}_{\text{Pointy-top}} = \frac{1}{3}\text{area}_{\text{base}} \cdot \text{height}$$

Surface area of pointy-top figures: The following formula gives you the surface area of a pyramid or cone.

$$\text{SA}_{\text{Pointy-top}} = \text{area}_{\text{base}} + \text{lateral area}_{\text{triangle(s)}}$$

The *lateral area* of a pointy-top figure is the area of the surface that connects the base to the peak (it's the area of everything but the base). Here's what this means for pyramids and cones:

✔ **The lateral area of a pyramid is made up of triangles.** Each lateral triangle has an area of $\frac{1}{2}$(base)(height). But you can't use the height of the pyramid for the height of its triangular faces, because the height of the pyramid doesn't run along its faces. So instead, you use the pyramid's *slant height,* which is just the ordinary altitude of the triangular faces. (The cursive letter ℓ indicates the slant height.) Figure 10-2 shows how height and slant height differ.

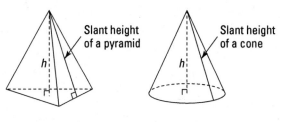

Figure 10-2: A pyramid and a cone with their heights and slant heights.

✔ **The lateral area of a cone is basically one triangle rolled into a cone shape.** The lateral area of a cone is one "triangle" that's been rolled into a cone shape like a snow-cone cup. (It's only sort of a triangle because when flattened out, it's actually a sector of a circle with a curved bottom side.) Its area is $\frac{1}{2}$(base)(slant height), just like one of the lateral triangles in a pyramid. The base of this "triangle" equals the circumference of the cone. So, the lateral area of a cone equals $\frac{1}{2}(2\pi r)\ell$, or $\pi r\ell$.

Given: A pyramid as shown

Diagonal $\overline{PT}$ has a length of 12

Find: 1. The pyramid's volume

2. The pyramid's surface area

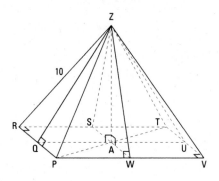

 The key to pyramid problems about volume and surface area is *right triangles*. Find them and then solve them with the Pythagorean Theorem or by using your knowledge of special right triangles.

1. **Find the pyramid's volume.**

To compute the volume of a pyramid, you need its height $(\overline{AZ})$ and the area of its square base, *PRTV*. You can get the height by solving right triangle $\triangle PZA$. The lateral edges of a pyramid are congruent; thus, the hypotenuse of $\triangle PZA$, $\overline{PZ}$, is congruent to $\overline{RZ}$, and so its length is also 10. $\overline{PA}$ is half of the diagonal of the base, so it's 6. Triangle *PZA* is thus a 3-4-5 triangle blown up to twice its size, namely a 6-8-10 triangle, so the height, $\overline{AZ}$, is 8 (or you can use the Pythagorean Theorem to get $\overline{AZ}$).

To get the area of square *PRTV*, you can, of course, first figure the length of its sides; but don't forget that a square is a kite, so you can use the kite area formula instead — that's the quickest way to get the area of a square if you know the length of a diagonal. Because the diagonals of a square are equal, both of them are 12, and you have what you need to use the kite area formula:

$$\text{Area}_{PRTV} = \frac{1}{2}d_1d_2$$
$$= \frac{1}{2}(12)(12)$$
$$= 72 \text{ units}^2$$

Now use the pointy-top volume formula:

$$\text{Vol}_{\text{Pyramid}} = \frac{1}{3}\text{area}_{\text{base}} \cdot \text{height}$$
$$= \frac{1}{3}(72)(8)$$
$$= 192 \text{ units}^3$$

2. Find the pyramid's surface area.

To use the pyramid surface area formula, you need the area of the base (which you got in part 1 of the problem) and the area of the triangular faces. To get the faces, you need the slant height, $\overline{ZW}$.

First, solve $\triangle PAW$. It's a 45°- 45°- 90° triangle with a hypotenuse $(\overline{PA})$ that's 6 units long; to get the legs, you divide the hypotenuse by $\sqrt{2}$ (or use the Pythagorean Theorem). $\frac{6}{\sqrt{2}} = 3\sqrt{2}$, so $\overline{PW}$ and $\overline{AW}$ both have a length of $3\sqrt{2}$. Now you can get $\overline{ZW}$ by using the Pythagorean Theorem with either of two right triangles, $\triangle PZW$ or $\triangle AZW$. Take your pick. How about $\triangle AZW$?

$$(ZW)^2 = (AZ)^2 + (AW)^2$$
$$= 8^2 + \left(3\sqrt{2}\right)^2$$
$$= 64 + 18$$
$$= 82$$
$$ZW = \sqrt{82}$$

Now you're all set to finish with the surface area formula. (One last fact you need is that $\overline{PV}$ is $6\sqrt{2}$ because, of course, it's twice as long as $\overline{PW}$.)

$$\text{SA}_{\text{Pyramid}} = \text{area}_{\text{base}} + \text{lateral area}_{\text{four triangles}}$$

$$= 72 + 4\left(\frac{1}{2}\text{base} \cdot \text{slant height}\right)$$

$$= 72 + 4\left(\frac{1}{2} \cdot 6\sqrt{2} \cdot \sqrt{82}\right)$$

$$= 72 + 12\sqrt{164}$$

$$= 72 + 24\sqrt{41}$$

$$\approx 225.7 \text{ units}^2$$

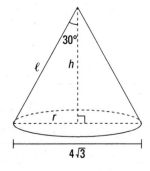

Given: Cone with base diameter of $4\sqrt{3}$
The angle between the cone's height and slant height is $30°$

Find:
1. The cone's volume
2. Its surface area

1. **Find the cone's volume.**

   To compute the cone's volume, you need its height and the radius of its base. The radius is, of course, half the diameter, so it's $2\sqrt{3}$. Then, because the height is perpendicular to the base, the triangle formed by the radius, the height, and the slant height is a $30°$-$60°$-$90°$ triangle. You can see that h is the long leg and r the short leg, so to get h, you multiply r by $\sqrt{3}$:

   $$h = \sqrt{3} \cdot 2\sqrt{3} = 6$$

You're ready to use the cone volume formula:

$$\text{Vol}_{\text{Cone}} = \frac{1}{3}\,\text{area}_{\text{base}} \cdot \text{height}$$

$$= \frac{1}{3}\pi r^2 \cdot h$$

$$= \frac{1}{3}\pi\left(2\sqrt{3}\right)^2 \cdot 6$$

$$= 24\pi$$

$$= 75.4 \text{ units}^3$$

2. **Find the cone's surface area.**

For the surface area, the only other thing you need is the slant height, ℓ. The slant height is the hypotenuse of the 30°- 60°- 90° triangle, so it's just twice the radius, which makes it $4\sqrt{3}$. Now plug everything into the cone surface area formula:

$$\text{SA}_{\text{Cone}} = \text{area}_{\text{base}} + \text{lateral area}_{\text{"triangle"}}$$

$$= \pi r^2 + \frac{1}{2}(\text{base})(\text{slant height})$$

$$= \pi r^2 + \frac{1}{2}(2\pi r)(\ell)$$

$$= \pi\left(2\sqrt{3}\right)^2 + \frac{1}{2}\left(2\pi \cdot 2\sqrt{3}\right)\left(4\sqrt{3}\right)$$

$$= 12\pi + 24\pi$$

$$= 36\pi$$

$$\approx 113.1 \text{ units}^2$$

Spheres

Volume and surface area of a sphere: Use the following formulas for the volume and surface area of a sphere.

- $\text{Vol}_{\text{Sphere}} = \frac{4}{3}\pi r^3$

- $\text{SA}_{\text{Sphere}} = 4\pi r^2$

Have a ball with the following sphere problem: What's the volume of a basketball in a box (a cube, of course) if the box has a surface area of 486 inches²?

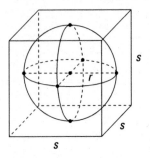

A cube (or any other ordinary box shape) is a special case of a prism, but you don't need to use the fancy-schmancy prism formula, because the surface area of a cube is simply made up of six congruent squares. Call the length of an edge of the cube s. The area of each side is therefore s^2. The cube has six faces, so its surface area is $6s^2$. Set this equal to the given surface area of 486 inches² and solve for s:

$$6s^2 = 486$$
$$s^2 = 81$$
$$s = 9 \text{ inches}$$

Thus, the edges of the cube are 9 inches, and because the basketball has the same width as the box it comes in, the diameter of the ball is also 9 inches; its radius is half of that, or 4.5 inches. Now you can finish by plugging 4.5 into the volume formula:

$$\text{Vol}_{\text{Sphere}} = \frac{4}{3}\pi r^3$$
$$= \frac{4}{3}\pi \cdot 4.5^3$$
$$= 121.5\pi$$
$$\approx 381.7 \text{ inches}^3$$

(By the way, this is slightly more than half the volume of the box, which is 9^3, or 729 inches³.)

Chapter 11

Coordinate Geometry

. . . .

. . . .

*I*n this chapter, you investigate the same sorts of things you see in previous chapters: perpendicular lines, right triangles, circles, perimeter, the diagonals of quadrilaterals, and so on. What's new about this chapter is that these familiar geometric objects are placed in the *x-y* coordinate system and then analyzed with algebra.

The Coordinate Plane

If you need a quick refresher about how the *x-y* coordinate system works, no worries. Check out Figure 11-1.

✔ Points are located within the coordinate plane with pairs of coordinates called *ordered pairs* — like (8, 6) or (–10, 3). The first number, the *x-coordinate*, tells you how far you go right or left; the second number, the *y-coordinate*, tells you how far you go up or down.

✔ Going counterclockwise from the upper-right-hand section of the coordinate plane are *quadrants* I, II, III, IV.

✔ The Pythagorean Theorem comes up a lot when you're using the coordinate system because when you go right and then up to plot a point (or left and then down, and so on), you're tracing along the legs of a right triangle; the segment connecting the *origin* (0, 0) to the point then becomes the hypotenuse of the right triangle. In Figure 11-1, you can see the 6-8-10 right triangle in quadrant I.

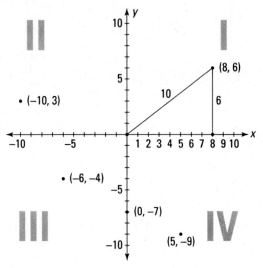

Figure 11-1: The *x-y* coordinate system.

Slope, Distance, and Midpoint

If you have two points in the coordinate plane, the three most basic questions you can ask about them are these:

- ✔ What's the distance between them?
- ✔ What's the location of the point halfway between them?
- ✔ How steep is the segment that connects the points?

The slope dope

You may have studied slope in an algebra class. But in case you've forgotten some of the concepts, here's a refresher.

Slope formula: The slope of a line containing two points — (x_1, y_1) and (x_2, y_2) — is given by the following formula (a line's slope is often represented by the letter *m*):

$$\text{Slope} = m = \frac{y_2 - y_1}{x_2 - x_1} = \frac{\text{rise}}{\text{run}}$$

The *rise* is the "up distance," and the *run* is the "across distance" shown in Figure 11-2. Note that you *rise up* but you *run across*, and also that "rise" rhymes with "*y*'s".

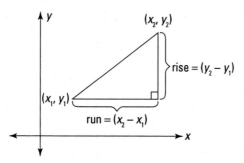

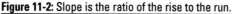

Figure 11-2: Slope is the ratio of the rise to the run.

Take a look at the following list and Figure 11-3, which show you that the slope of a line increases as the line gets steeper:

- ✔ A horizontal line has no steepness at all, so its slope is zero. Think about driving on a horizontal, flat road — the road has zero steepness or slope.

- ✔ A slightly inclined line might have a slope of, say, $\frac{1}{5}$.

- ✔ A line at a 45° angle has a slope of 1.

- ✔ A steeper line could have a slope of 5.

- ✔ A vertical line (the steepest line of all) sort of has an infinite slope, but math people say that its slope is *undefined*. (It's undefined because with a vertical line, you don't go across at all, and thus the *run* in $\frac{\text{rise}}{\text{run}}$ would be zero, and you can't divide by zero). Think about driving up a vertical road: It's impossible. And it's impossible to compute the slope of a vertical line.

- ✔ Lines that go up to the right have a *positive* slope. Going from left to right, lines with positive slopes go uphill.

- ✔ Lines that go down to the right have a *negative* slope. Going left to right on a negative slope, you go downhill.

Here are some pairs of lines with special slopes:

- ✔ **Slopes of parallel lines:** Parallel lines have the same slope.

- ✔ **Slopes of perpendicular lines:** The slopes of perpendicular lines are opposite reciprocals of each other, such as $\frac{7}{3}$ and $-\frac{3}{7}$ or -6 and $\frac{1}{6}$.

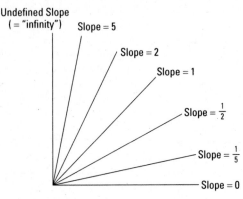

Figure 11-3: The slope tells you how steep a line is.

The distance formula

Ready for your next formula? Check it out below. And see Figure 11-4, which illustrates the formula.

Distance formula: The distance between two points (x_1, y_1) and (x_2, y_2) is given by the following formula:

$$\text{Distance} = \sqrt{(x_2 - x_1)^2 + (y_2 - y_1)^2}$$

The distance formula is simply the Pythagorean Theorem $(a^2 + b^2 = c^2)$ solved for the hypotenuse: $c = \sqrt{a^2 + b^2}$. See Figure 11-4 again. The legs of the right triangle (a and b under the square root symbol) have lengths equal to $(x_2 - x_1)$ and $(y_2 - y_1)$. Remember this connection, and if you forget the distance formula, you'll be able to solve a distance problem with the Pythagorean Theorem instead.

Don't mix up the slope formula with the distance formula. Both formulas involve the expressions $(x_2 - x_1)$ and $(y_2 - y_1)$. That's because the lengths of the legs of the right triangle in the distance formula are the same as the *rise* and the *run* from the slope formula. To keep the formulas straight, just focus on the fact that slope is a *ratio* of rise over run and that the distance formula *squares* things like the Pythagorean Theorem because it gives you the length of a hypotenuse.

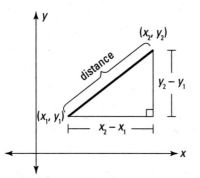

Figure 11-4: The distance between two points is also the length of the hypotenuse.

The midpoint formula

The midpoint formula takes the *average* of the x-coordinates of the segment's endpoints and the average of the y-coordinates of the endpoints.

Midpoint formula: To find the midpoint of a segment with endpoints at (x_1, y_1) and (x_2, y_2), use the following formula:

$$\text{Midpoint} = \left(\frac{x_1 + x_2}{2}, \frac{y_1 + y_2}{2} \right)$$

Trying out the formulas

Given: Quadrilateral *PQRS* as shown

Solve: 1. Show that *PQRS* is a rectangle

2. Find the perimeter of *PQRS*

3. Show that the diagonals of *PQRS* bisect each other, and find the point where they intersect

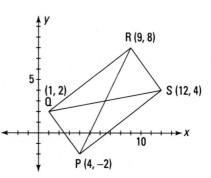

1. **Show that *PQRS* is a rectangle.**

   The easiest way to show that *PQRS* is a rectangle is to compute the slopes of its four sides and then use ideas about the slopes of parallel and perpendicular lines.

   $$\text{Slope} = \frac{y_2 - y_1}{x_2 - x_1}$$

   $$\text{Slope}_{\overline{QR}} = \frac{8-2}{9-1} = \frac{6}{8} = \frac{3}{4} \qquad \text{Slope}_{\overline{QP}} = \frac{2-(-2)}{1-4} = \frac{4}{-3} = -\frac{4}{3}$$

   $$\text{Slope}_{\overline{PS}} = \frac{4-(-2)}{12-4} = \frac{6}{8} = \frac{3}{4} \qquad \text{Slope}_{\overline{RS}} = \frac{8-4}{9-12} = \frac{4}{-3} = -\frac{4}{3}$$

   You can see that the four segments have a slope of either $\frac{3}{4}$ or $-\frac{4}{3}$ (which are opposite reciprocals). Thus, you can quickly see that, at each of the four vertices, perpendicular segments meet. All four vertices are therefore right angles, and a quadrilateral with four right angles is a rectangle.

2. **Find the perimeter of *PQRS*.**

   Because you now know that *PQRS* is a rectangle and that its opposite sides are therefore congruent, you need to compute the lengths of only two sides:

   $$\text{Distance} = \sqrt{(x_2 - x_1)^2 + (y_2 - y_1)^2}$$

   $$\text{Distance}_{P \text{ to } Q} = \sqrt{(1-4)^2 + [2-(-2)]^2}$$
   $$= \sqrt{(-3)^2 + 4^2} = 5$$

   $$\text{Distance}_{Q \text{ to } R} = \sqrt{(9-1)^2 + (8-2)^2}$$
   $$= \sqrt{8^2 + 6^2} = 10$$

   Now that you've got the length and width, you can easily compute the perimeter:

   $$\text{Perimeter}_{PQRS} = 2(\text{length}) + 2(\text{width})$$
   $$= 2(10) + 2(5) = 30$$

3. **Show that the diagonals of *PQRS* bisect each other, and find the point where they intersect.**

   If you know your rectangle properties, you know that the diagonals of *PQRS* must bisect each other. But another way to show this is with coordinate geometry. The term *bisect* in this problem should ring the *midpoint* bell.

$$\text{Midpoint} = \left(\frac{x_1 + x_2}{2}, \frac{y_1 + y_2}{2} \right)$$

$$\text{Midpoint}_{\overline{QS}} = \left(\frac{1+12}{2}, \frac{2+4}{2} \right) \qquad \text{Midpoint}_{\overline{PR}} = \left(\frac{4+9}{2}, \frac{-2+8}{2} \right)$$

$$= (6.5, 3) \qquad\qquad\qquad = (6.5, 3)$$

The fact that the two midpoints are the same shows that each diagonal goes through the midpoint of the other, and that, therefore, each diagonal bisects the other. Obviously, the diagonals cross at (6.5, 3).

Equations for Lines and Circles

If you've already taken algebra, you've probably dealt with graphing lines in the coordinate system. Graphing circles may be new for you, but you'll soon see that there's nothing to it.

Line equations

Here are the basic forms for equations of lines:

✔ **Slope-intercept form:** Use this form when you know (or can easily find) a line's slope and its *y*-intercept.

$y = mx + b,$
where *m* is the slope and *b* is the *y*-intercept $(0, b)$

✔ **Point-slope form:** Use this form when you don't know a line's *y*-intercept but you do know the coordinates of a point on the line; you also need the line's slope.

$y - y_1 = m(x - x_1),$
where *m* is the slope and (x_1, y_1) is a point on the line

✔ **Horizontal line:** This form is used for horizontal lines.

$y = b,$
where *b* is the *y*-intercept

Note that every point along a horizontal line has the same *y*-coordinate, namely the number you put into *b*. (This equation is a special case of $y = mx + b$ where $m = 0$.)

✔ **Vertical line:** Here's the equation for a vertical line.

$x = a$,
where a is the x-intercept

Every point along a vertical line has the same x-coordinate, namely the number you put into a.

Because a horizontal line is parallel to the x-axis, you might think that the equation of a horizontal line would be $x = a$. And you might figure that the equation for a vertical line would be $y = b$ because a vertical line is parallel to the y-axis. But it's the other way around.

The circle equation

✔ **Circle centered at the origin, (0, 0):**

$x^2 + y^2 = r^2$

✔ **Circle centered at any point (h, k):**

$(x - h)^2 + (y - k)^2 = r^2$,

where (h, k) is the center of the circle and r is its radius

Ready for a circle problem? Here you go:

Given: Circle C has its center at $(4, 6)$ and is tangent to a line at $(1, 2)$

Find: 1. The equation of the circle

2. The circle's x- and y- intercepts

3. The equation of the tangent line

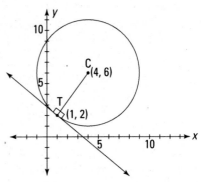

1. Find the equation of the circle.

All you need for the equation of a circle is its center (you know it) and its radius. The radius of the circle is just the distance from its center to any point on the circle, but since the point of tangency is given, that's the easiest point to use. To wit —

$$\text{Distance}_{C \text{ to } T} = \sqrt{(4-1)^2 + (6-2)^2}$$
$$= \sqrt{3^2 + 4^2} = 5$$

Now you finish by plugging the center coordinates and the radius into the general circle equation:

$$(x-h)^2 + (y-k)^2 = r^2$$
$$(x-4)^2 + (y-6)^2 = 5^2$$

2. **Find the circle's *x*- and *y*-intercepts.**

   To find the *x*-intercepts for any equation, you just plug in 0 for *y* and solve for *x*:

   $$(x-4)^2 + (y-6)^2 = 5^2$$
   $$(x-4)^2 + (0-6)^2 = 5^2$$
   $$(x-4)^2 + 36 = 25$$
   $$(x-4)^2 = -11$$

   You can't square something and get a negative number, so this equation has no solution; therefore, the circle has no *x*-intercepts. (I realize that you can just look at the figure and see that the circle doesn't intersect the *x*-axis, but I wanted to show you how the math confirms this.)

   To find the *y*-intercepts, plug in 0 for *x* and solve:

   $$(0-4)^2 + (y-6)^2 = 5^2$$
   $$16 + (y-6)^2 = 25$$
   $$(y-6)^2 = 9$$
   $$y-6 = \pm\sqrt{9}$$
   $$y = \pm 3 + 6$$
   $$= 3 \text{ or } 9$$

   Thus, the circle's *y*-intercepts are (0, 3) and (0, 9).

3. **Find the equation of the tangent line.**

For the equation of a line, you need a point (you have it) and the line's slope. Since a tangent line is perpendicular to the radius drawn to the point of tangency, you just compute the slope of the radius, and then the opposite reciprocal of that is the slope of the tangent line.

$$\text{Slope} = \frac{y_2 - y_1}{x_2 - x_1}$$

$$\text{Slope}_{\text{Radius } \overline{CT}} = \frac{6 - 2}{4 - 1} = \frac{4}{3}$$

Therefore,

$$\text{Slope}_{\text{Tangent line}} = -\frac{3}{4}$$

Now you plug this slope and the coordinates of the point of tangency into the point-slope form for the equation of a line:

$$y - y_1 = m(x - x_1)$$

$$y - 2 = -\frac{3}{4}(x - 1)$$

Chapter 12

Ten Big Reasons to Use in Proofs

*H*ere's the top ten list of definitions, postulates, and theorems that you should absotively, posilutely know how to use in the reason column of geometry proofs.

The Reflexive Property

The *Reflexive Property* says that any segment or angle is congruent to itself. You often use the Reflexive Property, which I introduce in Chapter 5, when you're trying to prove triangles congruent or similar. Be careful to notice all shared segments and shared angles in proof diagrams. Shared segments are usually pretty easy to spot, but people sometimes fail to notice shared angles.

Vertical Angles Are Congruent

The vertical angle theorem isn't hard to use, as long as you spot the vertical angles (see Chapters 1 and 2). Remember — everywhere you see two lines that come together to make an X, you have *two* pairs of congruent vertical angles (the ones on the top and bottom of the X and the ones on the left and right sides of the X).

The Parallel-Line Theorems

There are ten parallel-line theorems that involve a pair of parallel lines and a transversal (which intersects the parallel lines). Look back to Figure 6-1. Five of the theorems use parallel lines to show that angles are congruent or supplementary; the reverse of these five use congruent or supplementary angles to show that lines are parallel.

If lines are parallel, then . . .

  ✔ Alternate interior angles, like $\angle 4$ and $\angle 5$, are congruent.

  ✔ Alternate exterior angles, like $\angle 1$ and $\angle 8$, are congruent.

  ✔ Corresponding angles, like $\angle 3$ and $\angle 7$, are congruent.

  ✔ Same-side interior angles, like $\angle 4$ and $\angle 6$, are supplementary.

  ✔ Same-side exterior angles, like $\angle 1$ and $\angle 7$, are supplementary.

And here are the ways to prove lines parallel:

  ✔ If alternate interior angles are congruent, then lines are parallel.

  ✔ If alternate exterior angles are congruent, then lines are parallel.

  ✔ If corresponding angles are congruent, then lines are parallel.

  ✔ If same-side interior angles are supplementary, then lines are parallel.

  ✔ If same-side exterior angles are supplementary, then lines are parallel.

Two Points Determine a Line

Not much to be said here — whenever you have two points, you can draw a line through them. Two points *determine* a line because only one particular line can go through both points. You use this postulate in proofs whenever you need to draw an auxiliary line on the diagram (see Chapter 6).

All Radii Are Congruent

Whenever you have a circle in your proof diagram, you should think about the all-radii-are-congruent theorem (and then mark all radii congruent) before doing anything else. I bet that just about every circle proof you see will use congruent radii somewhere in the solution. (And you'll often have to use the postulate in the preceding section to draw in more radii.) I discuss this theorem in Chapter 9.

If Sides, Then Angles

Isosceles triangles have two congruent sides and two congruent base angles. The if-sides-then-angles theorem says that if two sides of a triangle are congruent, then the angles opposite those sides are congruent (look back to Figure 5-7). Do not fail to spot this! When you have a proof diagram with triangles in it, always check to see whether any triangle looks like it has two congruent sides. For more information, flip to Chapter 5.

If Angles, Then Sides

The if-angles-then-sides theorem says that if two angles of a triangle are congruent, then the sides opposite those angles are congruent (check out Figure 5-8). Yes, this theorem is the converse of the if-sides-then-angles theorem, so you may be wondering why I didn't put this theorem in the preceding section. Well, these two isosceles triangle theorems are so important that each deserves its own section.

Triangle Congruence

Here are the five ways to prove triangles congruent (see Chapter 5 for details).

- ✔ **SSS (side-side-side):** If the three sides of one triangle are congruent to the three sides of another triangle, then the triangles are congruent.

- ✔ **SAS (side-angle-side):** If two sides and the included angle of one triangle are congruent to two sides and the included angle of another triangle, then the triangles are congruent.

✓ **ASA (angle-side-angle):** If two angles and the included side of one triangle are congruent to two angles and the included side of another triangle, then the triangles are congruent.

✓ **AAS (angle-angle-side):** If two angles and a nonincluded side of one triangle are congruent to two angles and a nonincluded side of another triangle, then the triangles are congruent.

✓ **HLR (hypotenuse-leg-right angle):** If the hypotenuse and a leg of one right triangle are congruent to the hypotenuse and a leg of another right triangle, then the triangles are congruent.

CPCTC

CPCTC stands for *corresponding parts of congruent triangles are congruent*. It has the feel of a theorem, but it's really just the definition of congruent triangles. When doing a proof, after proving triangles congruent, you use CPCTC on the next line to show that some parts of those triangles are congruent. CPCTC makes its debut in Chapter 5.

Triangle Similarity

Here are the three ways to prove triangles similar — that is, to show they have the same shape (see Chapter 8 for details).

✓ **AA (angle-angle):** If two angles of one triangle are congruent to two angles of another triangle, then the triangles are similar.

✓ **SSS~ (side-side-side similar):** If the ratios of the three pairs of corresponding sides of two triangles are equal, then the triangles are similar.

✓ **SAS~ (side-angle-side similar):** If the ratios of two pairs of corresponding sides of two triangles are equal and the included angles are congruent, then the triangles are similar.

Index